IRELAND: Some Problems of a Developing Economy

Also edited by A. A. Tait and J. A. Bristow
ECONOMIC POLICY IN IRELAND

IRELAND

Some Problems of a Developing Economy

Edited by
A. A. TAIT AND J. A. BRISTOW

GILL AND MACMILLAN
Dublin

BARNES AND NOBLE, PUBLISHERS
New York

First published in Ireland 1972 by
GILL AND MACMILLAN LTD
2 Belvedere Place
Dublin 1

in London through association with the
MACMILLAN
Group of Publishing Companies

First published in the United States of America, 1972 by
BARNES & NOBLE, PUBLISHERS
New York, New York

Gill & Macmillan SBN: 7171 0500 8
Barnes & Noble ISBN 389 04452 9

Jacket design by Cor Klaasen

Printing History
10 9 8 7 6 5 4 3 2 1

Printed and bound in the Republic of Ireland by the Book Printing Division of
Smurfit Print and Packaging Limited, Dublin

Contents

Preface   vii

I   *Fiscal Policy and Demand Management in Ireland, 1960–70*   LOUDEN RYAN   1

II   *Some Issues of Irish Debt Policy*   ALAN A. TAIT   38

III   *The Flexibility of Irish Taxes on Incomes*   L. K. LENNAN   68

IV   *Hidden Subsidies in Irish Electricity Supply*   J. A. BRISTOW   98

V   *The Elasticity of Demand for Irish Exports*   M. J. HARRISON   124

VI   *The Distribution of Personal Wealth in Ireland*   PATRICK M. LYONS   159

VII   *The Brain Drain Question*   M. O'DONOGHUE   186

VIII   *Wage Inflation and Wage Leadership in Ireland, 1954–69*   C. MULVEY AND J. TREVITHICK   204

Index   235

PREFACE

THIS is not a comprehensive treatise on Irish economic development. Each chapter deals with a question of importance to policy-makers in a developmental situation; thus, these are pieces in an incomplete jig-saw and a major function of this preface is to indicate how these pieces fit into the overall picture of Irish economic development.

Ireland provides a good illustration of the difficulties of dividing economies into neat parcels labelled 'developed' and 'developing'. It has a 'European' level of per capita income and by world standards well-developed fiscal and monetary systems. Furthermore, it is currently experiencing the problems of cost inflation now so familiar in, say, Britain and the U.S.A. In these and other respects it is developed. At the same time it is an ex-colony whose economy would, until recently, be justifiably called agricultural. Attempts to transform it into an industrial economy have shown significant success only in the past fifteen years. It therefore faces many of the major structural problems familiar in developing countries. The difficulties of effecting a rapid and fundamental change in economic structure have implications for practically every aspect of economic policy, and so Ireland is most interestingly thought of as a developing country: hence the title of this book.

The first four chapters deal directly with public sector operations and reflect the problems of an active government intent on promoting economic growth. Since the mid-1960s Ireland has had the 'luxury' of having to cope with recurrent inflationary pressures which in turn pose problems for the balance of payments. Chapters I and III discuss two aspects of this stabilisation problem. A government which claims to be Keynesian has been inhibited in tackling the question of demand management by an adherence to conservative budgetary concepts and procedures. This conflict is dealt with in Chapter I. When, however, political, institutional or economic factors make discretionary fiscal policy difficult, some consolation can be derived from the existence of built-in stabilisers. Chapter III discusses the yield flexibility of Irish income taxes in this context.

The temptations and difficulties of financing government expenditure by borrowing create problems which many countries other than Ireland experience. But Ireland has the second largest national debt per capita in the world, and this adds piquancy to the discussion found in Chapter II.

Many developing countries have sought some of the solutions to the problems of transforming a primarily agrarian economy into an industrial one in extensive state participation in the industrial sector. Ireland is no exception. But allocative and distributive questions arise when the pricing and investment policies of public enterprises are required to accommodate multiple objectives. Chapter IV examines these questions in relation to a major nationalised industry.

The next four chapters are concerned with the private sector, but with aspects of that sector which are of particular interest in the context of economic development. Thus, Ireland's development prospects are vitally influenced by export performance and of particular importance is the possible role of exchange-rate adjustments. Chapter V examines some of the difficulties in assessing the effects of exchange rate changes.

Another problem arising from economic development is that of the distribution of income and wealth. Chapter VI uses a well-known and accepted method to calculate for the first time the distribution of wealth in Ireland. The results pose some interesting policy problems and highlight some of the peculiar characteristics of the Irish economy.

In recent years many developing countries have been concerned about the emigration of educated manpower. Again Ireland is no exception. Perhaps because of the inevitable social dimensions of the problem, the 'brain drain' is often discussed too emotionally: Chapter VII presents the economic issues in a more dispassionate manner and indicates the direction future research should take if the effects on the Irish economy of this emigration are to be more fully evaluated.

Finally, Chapter VIII looks at one aspect of the cost inflation currently of great concern to Irish policy-makers, an aspect arising from a fascinating feature of the Irish economic scene— the wage round. Wage rounds (periodic, generalised wage increases) have been a characteristic of post-war Ireland and have contributed significantly to a rate of increase in unit labour costs exceeding that in those countries which form Ireland's major export markets. It may be that wage leadership is crucial here, and Chapter VIII attempts to identify this phenomenon.

We who live in Ireland often think it is unique: or when depressed by some of our seemingly intractable problems we may hope, for the sake of the rest of the world, that it is. But of course it is not. All our economic problems, and more specifically all those discussed here, arise to a greater or lesser degree in most countries, and especially all those wrestling with economic development. For this reason we hope that this book will provide something of interest to those everywhere who have a professional or other concern for the problems of economic development.

The authors are all on the staff of the economics department at Trinity College, Dublin, with the exception of Mr Lennan who is a graduate student.

A.A.T.
J.A.B.

Dublin,
February, 1971.

CHAPTER I

Fiscal Policy and Demand Management in Ireland 1960-70[1]

LOUDEN RYAN

THIS chapter discusses fiscal policy and the management of the Irish economy during the years 1960–70. The role of fiscal policy in demand management, as most economists would probably nowadays see it, is described in the first section. In the second section the annual current and capital budgets for the years 1960–70 are examined. In discussing the annual budgets, the emphasis is on their role and relevance as seen by the successive Ministers for Finance who introduced them. In the final section, the main conclusions are summarised.

ROLE OF FISCAL POLICY

In Ireland, as in other countries, the major economic objectives are high and steady growth, full employment, an acceptable external position, stable prices, the efficient use of all productive resources and equity in the distribution of income and wealth. Fiscal policy, working through the annual current and capital budgets, is a major instrument by which these aims can be pursued. By raising investment, providing infrastructure and improving education, the capital budget lays the foundation for their achievement in the longer term. By maintaining the right relationship between aggregate demand and the possibilities of meeting it without excessive inflation or insupportable external deficits, both the current and capital budgets can assist in the short run by maintaining the conditions necessary for the achievement of these objectives. It is these short-term functions of fiscal

[1] The author is indebted to Dr T. K. Whitaker, Mr C. H. Murray, Professor P. Lynch, Dr Kieran Kennedy, the editors, Mr C. Kavanagh and Mr T. Hoare for reading an earlier draft and for their many helpful comments and suggestions. The chapter has benefited a great deal from their assistance. The author remains responsible for its deficiencies.

policy which are emphasised in the context of demand management.

Fiscal policy can influence the balance between aggregate demand and supply in the economy through changes in government expenditures and revenues. Taken by itself, an increase in government expenditures will raise aggregate demand, and the extent of the rise will depend on the nature of the expenditures. An additional £1 million spent on social welfare payments, for example, will raise the incomes of the recipients by that amount. The greater part of the additional incomes will be spent on consumer goods and services, thus generating further incomes for those who produced and sold them in Ireland, for government itself (through the taxes levied on the goods and services bought) or for people outside Ireland (to the extent that the goods were imported or had an import content). An additional £1 million spent on new housing accrues as wages and salaries to the workers employed directly or indirectly on it, to contractors and builders' providers as additional profits, to the government (as additional tax revenue) and as incomes to externals (to the extent that imported goods and services are used). Some (probably the greater) part of the new incomes thus generated by the additional government expenditures will be spent, generating further increases in incomes, and so on in a diminishing series.

An increase in the flow of money to the government will, other things remaining equal, affect the level of aggregate demand, generally tending to lower it. The size of the decrease in aggregate demand will depend on the way in which the money is raised. If the additional funds are obtained from increases in the rates of direct or indirect taxation, or increases in the yield from taxation at the existing rates, purchasing power is transferred from the taxpayers to the government. As a result of this transfer, taxpayers will reduce both their (real) spending on consumption goods and services and their saving.[2] The reduction in their spending will generate fewer (or smaller) incomes for Irish

---

[2] The effects may vary, depending on whether the transfer is effected by increases in direct or indirect taxes. A recent study has concluded that increased direct taxes are likely to be unfavourable to savings while increased indirect taxes tend to raise the savings ratio (that is, tend to reduce consumption). See Kieran A. Kennedy and Brendan R. Dowling, 'The Determinants of Personal Savings in Ireland: An Econometric Inquiry', *Economic and Social Review* II, 1, (October 1970), 19–51.

residents (to the extent that fewer Irish goods are now being sold), smaller revenue for the government (through reduced tax receipts), or for externals (to the extent that the impact of the reduced spending falls on imports).

If the additional funds are raised by borrowing from the Irish non-bank public,[3] the effects are more difficult to assess. Like taxation, borrowing from the non-bank public involves a transfer of command over resources from the private sector to the government. If interest rates are responsive to changes in the supply of and demand for domestic securities, they will tend to rise. The rise in interest rates is unlikely to have any significant and direct effect on current savings. Its main effect will be to reallocate the available savings between the public and private sectors—to reduce private sector borrowing and thus help to accommodate the new government borrowing. It may also increase the 'voluntary' capital inflow. In addition, wealth effects may follow from the higher interest rates: the associated fall in the value of security portfolios may lead to some reduction in consumption and a rise in saving; and there may be some shift in the asset portfolios of the private sector from more liquid towards less liquid assets. If the banking system adhered to fixed cash and liquidity ratios, this rearrangement in asset portfolios could lead to some reduction in bank lending (because bank deposits would fall as a result of the shift towards less liquid assets), with some consequential reductions in private expenditures.

Where interest rates are not responsive to changes in the domestic capital market (as is largely the case in Ireland), the government borrowing is met by some form of rationing: for example, the government usually announces the size and terms of a national loan well in advance of the date when subscriptions are invited, so that institutional and other lenders can take the government's requirements into account when planning the additions to their portfolios. To the extent that the loan is taken up, that much less is available for the private sector. Where there is no change in interest rates, there is no wealth effect on consumption and no change in private asset portfolios.

[3] It is assumed that the sums lent by the Irish non-bank public come from current saving. If the monies lent came from the sale of external assets by non-banks, the effects would be similar to Exchequer borrowing from the banking system or from abroad.

Generally, if there is no change in the money supply,[4] increased government borrowing from the non-bank public may reduce aggregate spending by the private sector by as much or more than the raising of the same amount by additional taxation. With increased direct or indirect taxes, there will be some reduction in private saving, so that private consumption spending will fall by less than the tax proceeds. With increased borrowing from the non-bank public with no change in the money supply, the fall in private (and mainly investment) expenditures will be of the same order as the additional funds borrowed, except in the unlikely event of current savings rising in response to higher interest rates or of the velocity of circulation rising in response to the squeeze on funds for use in the private sector. To the extent that borrowing from the non-bank public reduces private investment, the growth in output may be smaller in the later periods in which the government will have to raise additional funds to pay interest and repay the principal of its initial borrowings.[5]

The final possibility is that the additional funds required by the government are raised by borrowing from the banking system or from abroad. If the monetary authorities have fixed a ceiling for new credit creation (increase in the money supply), additional government borrowing from the banking system will tend to reduce the amount of new credit available to private borrowers and private spending will be less than it otherwise would have been. If the liquidity ratios of the banks are flexible downwards (as they were in the 1960s), and if there is no ceiling for new credit creation, or if the ceiling is tacitly raised to accommodate the additional requirements of the government, there will be no reduction in private spending. Where the additional requirements of the government are met from external borrowing, or by borrowing from the Central Bank, there will be no reduction in private spending. Indeed, the latter may rise if the banks adhere strictly to fixed cash and liquidity ratios, because foreign

---

[4] For a discussion of what happens if this asumption is relaxed, see next paragraph.

[5] This assumes a given level of government expenditures. If the given level of spending could not be sustained without borrowing then this conclusion would, of course, have to be modified.

borrowing by adding to the cash and liquidity base would make possible an increase in domestic bank lending.[6]

The argument to this stage may be summarised briefly. An increase in government expenditures will raise aggregate demand in the economy, generally by some multiple of itself. If this increase is financed (for example) by increased direct taxation, there will be some offsetting reduction in aggregate demand as private (mainly consumption) expenditures are reduced, as a result of the fall in private disposable income. The expansionary effects of an increase in government expenditures will normally exceed the contractionary effects of the increase in taxation which finances it; this occurs because to some extent the higher taxes are 'paid' from income that would otherwise have been saved, and the government spends all of the funds so raised.[7] The net expansionary effects will be greater to the extent that taxpayers succeed in passing on the increases in direct or indirect taxes by increases in their money incomes. If the increased expenditures are financed by increased borrowing from the Irish non-bank public, there will also be offsetting reductions in aggregate demand as private (mainly investment) expenditures decline.[8] Where the increased expenditures are financed by borrowing from the banking system or from abroad, and where monetary policy remains passive, the rise in aggregate demand will be determined by the rise in government expenditures since there will be no offsetting reductions in private expenditures. The government must at least raise the money to finance its expenditures[9]: by altering the level of its expenditures and/or the balance between the methods by which they are financed, the government

[6] The argument in this paragraph would need to be modified to the extent that funds borrowed from the banks or from abroad are spent externally. If they were all so spent there would be no internal multiplier effect.

[7] This is the concept of the 'balanced budget multiplier'. See, for example, R. A. Musgrave, *The Theory of Public Finance,* New York 1959, Chapters 18 and 19.

[8] Unless the subscriptions to new government loans came from 'dishoarding'. Until the beginning of the 1960s, the Associated Banks may have been underlent, so that the use of bank deposits to buy new government debt was probably not associated with any offsetting reduction in the funds available to the private sector. This has become increasingly less true over the last ten years or so. The statement in the text is valid for recent years.

[9] It may, on occasions, raise more from taxation than is required to meet its expenditures, the excess being applied to net debt redemption.

can change the level of aggregate demand in the economy, to bring it nearer to what is deemed appropriate.

In the previous paragraphs, no distinction was drawn between different categories of government expenditures (such as current and capital), and no reference was made to the appropriateness of financing any particular category in any particular way (for example, financing capital expenditure by borrowing and current expenditure by taxation). From the point of view of demand management, these particular distinctions and conventions are largely irrelevant. The starting point, when changes in fiscal policy are being planned, must be the existing position. Fiscal policy is applied by marginal changes in individual categories of expenditures or in the total, by marginal changes in taxation and in non-bank or bank and external borrowing, and by marginal shifts in their relative importance. Conventions which relate particular categories of expenditure to particular sources of funds (for example, that current expenditures and current tax revenues should be in balance) are equally irrelevant in this context. Changes in fiscal policy should be determined by reference to the state of the economy rather than by reference to the state of any part or all of the government accounts.

It was stated earlier that the aim of fiscal policy is to achieve an appropriate level of aggregate demand. In its simplest terms, the appropriate level of demand is that which would ensure some 'desirable' degree of utilisation of domestic productive capacity while maintaining reasonable price stability and keeping the balance of payments in an 'acceptable' or 'sustainable' position. Any estimate of this level of aggregate demand must be based on economic forecasts. These forecasts are generally related to the year ahead and are made to fit in with the timing of the annual budget(s). The starting point is usually estimates of potential capacity and of private income and consumption expenditure. To the latter, there are added estimates of exports, private investment and government expenditure. The first of these is a datum since no single country has in the short run much influence over the demand for its exports. The second—private fixed investment— may be largely determined for the year or so ahead by decisions already made. Government expenditures, both current and capital, are also fixed at some *minimum* level by past decisions, and can only be reduced below that level at the cost of injustice or inefficiency. This means that increases in government expenditures and in private consumption tend to become the only two variables on which fiscal policy can operate, and changes in taxation virtually the sole instrument by which the latter can be achieved.[10] If the forecast aggregate demand is above the level deemed appropriate by reference to the expected growth in capacity, successful demand management requires that planned or desired increases in government expenditure must be abandoned or cut, and/or private consumption reduced by increased taxation.

The implication in the previous paragraph that the appropiate level of aggregate demand can be uniquely identified is an over-simplification. Even if the likelihood of errors in forecasting is ignored, this would be true only if the sole objective of policy was some particular and consistent combination of capacity utilisation and external position. In fact, all governments simultaneously pursue a number of different objectives: full employment, growth, an acceptable balance of payments position, reasonable price stability, efficient use of productive resources, and equity in the distribution of income and wealth. Very often, policies to realise more fully any one of these objectives mean that one or more of the others will be achieved to a smaller extent. Thus, fuller employment may accelerate the rise in prices; an acceptable balance of payments deficit may be achieved only at the expense of slower growth, or more efficient use of resources may mean higher unemployment. Such 'trade-offs' between objectives are a matter for political decision and judgement.

Once the relative importance of the different objectives is decided, the appropriate level and composition of aggregate demand are largely determined, and the task of fiscal policy is to contribute towards their achievement. The desired level of aggregate demand can be sought as described in the preceding paragraphs. The desired composition can be achieved by the appropriate choice of fiscal instruments. For example, if more weight is attached to growth than external balance, the appropriate level of aggregate demand will be higher, increases in current rather than capital expenditures will be curbed, and greater emphasis may be placed on increases in indirect taxes

---

[10] Given the level of total government expenditure and of non-bank borrowing changes in tax revenue mean equal and opposite changes in the residual borrowing requirement (i.e. in borrowing from the banking system and from abroad).

rather than in direct taxes to restrain private spending. If more importance is attached to price stability than to full employment, the appropriate level of aggregate demand will be lower, and increases in direct taxes may be preferred to increases in indirect taxes.

Finally, it must be emphasised that the use of fiscal policy for demand management cannot be determined by any mechanical rule or formula. Different government expenditures will have different effects on the demand for goods and services produced in Ireland and the fraction that reaches households as income will vary from one kind of expenditure to another. Changes in income tax may have different effects on private expenditures from a change in turnover tax. To the extent that profits are affected by the tax changes, investment and the spending of firms' owners could be reduced. If incomes and prices are pushed up following the tax increases, the ultimate location of the burden would in most circumstances be indeterminate. Again, the effects of a deficit or surplus on the current budget will depend on how the deficit is financed or the surplus used. Moreover, the application of fiscal policy is complicated by inescapable time-lags in diagnosing whether (for example) a current budget deficit or surplus would be appropriate, and in deciding which fiscal measures to apply. Even after fiscal measures have been introduced, some time will elapse before their effects on aggregate demand are felt, and this time-lag will vary from one fiscal measure to another. These problems are not, of course, peculiar to the use of fiscal policy for the purpose of demand management—they would arise whatever objective was rationally pursued. What matters is what fiscal policy is trying to do.

IRISH BUDGETS 1960–70

In this section, the aim is to identify from the annual budget statements what might be called the philosophy of fiscal policy— its aims, rules and conventions. This approach is not wholly satisfactory. Fiscal policy should be judged by the effect on the economy of the fiscal measures actually applied, rather than from statements of its intentions or annual descriptions of its role as it is currently seen. In most countries, there exist conventions (such as that the annual current budget should be balanced) which both inhibit desirable changes and prevent undesirable changes by the discipline they impose. The problem of effecting change while incurring minimum risks is often solved by describing the new developments in the language, or within the framework, of the old conventions, so that statements of intent may not accurately describe what is in fact done. Moreover, ministerial statements about fiscal policy are often in the nature of exhortations to the public to consume less or to invest more in national loans, and should not be taken at their face value as statements of a budgetary philosophy.

To make sure of doing justice to fiscal (or monetary or any other) policy, judgements should be based mainly on what was actually done and on the effects that followed from it. The effects of changes in fiscal policy can be measured only by comparing the economy after their application with the state the economy would have been in if they had not been applied. It is not possible to make this comparision without a detailed and sophisticated model of the economy. As yet, sufficiently detailed models are not available for any economy to permit an assessment of the full effects of fiscal policy. For some countries (but not for Ireland) there are models which are sufficiently good for an attempt to measure the effects of fiscal policy on aggregate demand.[11] Since the effects cannot be measured at this stage for Ireland, an attempt is made in this section to judge fiscal policy by its intentions and actions.

In Ireland there are two parts to the annual budget, the current and the capital. Table 1·1 gives the estimated and actual figures for the current budget for each head of revenue and expenditure for each of the financial years 1960–1 to 1969–70, with the estimated figures for 1970–1. An alternative and more detailed classification of current government expenditures is given in Table 1·2. At the bottom of this table, current government expenditures are given as a percentage of the gross national product at current market prices. In all cases, the figures are taken from the tables published with the Annual Budget Statements.

[11] For attempts to measure the effects of fiscal policy on aggregate demand, see Bent Hansen (assisted by Wayne W. Snyder), *Fiscal Policy in Seven Countries 1955–1965,* O.E.C.D. Paris, March 1969.

	1960–1		1961–2		1962–3		1963–4		1964–5	
	Est.	*Act.*	*Est.*	*Act.*	*Est.*	*Act.*	*Est.*	*Act.*	*Est.*	*Ac*
Revenue										
1. Tax revenue (excl. 2 below)	106·72	108·43	111·64	119·22	129·12	128·76	144·63[5]	146·80[5]	173·01	17€
2. Motor vehicle duties	6·00	6·46	6·65	6·93	7·15	7·40	7·70	8·24	9·00	٤
3. Non-tax revenue Post Office	9·65	9·70	10·30	10·50	11·64	11·44	12·15	12·10	15·10	14
Miscellaneous	14·26	14·25	14·98	15·03	15·64	15·88	17·09	17·28	18·02	18
4. Deficit	—	0·73	—	0·71	—	4·85	—	2·22	—	4
Total	136·63	139·57	143·57	152·39	163·55	168·33	181·57	186·64	215·13	223
Expenditure										
1. Central fund services (excl. 2 below)	24·65	24·65	26·97	27·23	29·61	28·83	32·32	32·05	35·54	36
2. Payments to road fund	6·00	6·46	6·65	6·93	7·15	7·40	7·70	8·24	9·00	8
3. Supply services (non-capital)	108·48	108·46	112·95	118·23	129·19	132·10	143·55	146·35	170·59	178
4. Savings and over-estimation (net *deduction* from expenditure)	2·50	—	3·00	—	2·40	—	2·00	—	—	—
5. Surplus	—	—	—	—	—	—	—	—	—	—
	136·63	139·57	143·57	152·39	163·55	168·33	181·57	186·64	215·13	223

Notes: (1) The estimate and outcome for 1966–7 take account of adjustmen

(2) The figures in this column are the revised estimates for 1965–6 ma

(3) Estimates as at 23 April, 1968, including tax and expenditure chang

(4) Estimates taking account of the November Budget, 1968.

(5) Includes £2 million revenue balances.

Comparison between estimates and outcome. (£ million)

1965–6		1966–7		1967–8		1968–9			1969–70		1970–1
Est.	*Act.*[2]	*Est.*	*Act.*	*Est.*	*Act.*	*Est.*[3]	*Est.*[4]	*Act.*	*Est.*	*Act.*	*Est.*
196·81	193·92	217·44	221·55	238·51	247·98	270·23	281·03	282·30	322·32	337·57	393·70
9·40	9·40	11·05	10·43	11·50	11·60	12·16	12·16	12·70	13·10	13·41	14·83
16·20	16·40	17·40	18·33	20·50	20·50 }	50·30	22·90	21·70	27·10	27·35 }	66·36
20·40	20·62	21·84	22·53	24·71	25·33 }		28·23	28·78	31·80	32·68 }	
—	8·00	—	—	—	0·21	—	7·12	8·37	—	0·54	—
242·81	248·34	267·73	272·84	295·22	305·62	332·69	351·44	353·85	393·22	411·55	474·89
40·20	41·55	46·90	47·84	54·58	54·16	64·00	65·20	65·11	76·45	76·98	88·36
9·40	9·40	9·75	9·22	10·00	10·22	10·70	10·70	11·16	11·50	12·00	12·85
197·21	197·39	211·08	214·99	234·64	241·24	261·99	275·54	277·58	307·27	322·57	373·68
4·00	—	—	—	4·00	—	4·00	—	—	2·00	—	—
—	—	—	0·79	—	—	—	—	—	—	—	—
242·81	248·34	267·73	272·84	295·22	305·62	332·69	351·44	353·85	393·22	411·55	474·89

made in the March and June Budgets, 1966.

at 9 March, 1966.

made at that time.

TABLE 1.2

Main Heads of Government Current Expenditure, 1960–71.

£ million

	1960-1	1961-2	1962-3	1963-4	1964-5	1965-6	1966-7	1967-8	1968-9	1969-70	
										Prov.	F
Service of Public Debt	28·4	31·1	34·4	38·2	42·8	49·0	56·5	63·7	75·9	88·8	
Social Services	50·4	51·9	57·5	63·2	75·2	84·2	93·1	102·9	118·7	144·2	
Social Welfare	26·1	25·7	27·9	30·9	34·9	38·7	43·0	45·1	49·1	59·2	
Education	15·6	16·6	18·9	20·6	26·1	29·6	31·3	35·8	44·0	52·9	
Health	8·8	9·6	10·7	11·7	14·2	15·9	18·9	22·1	25·7	32·1	
Economic Services	24·9	30·6	35·0	39·2	47·9	55·2	59·9	75·2	84·4	97·5	1
Agriculture	14·1	18·9	22·3	24·0	30·0	35·8	40·8	53·3	60·0	70·8	
Industry	1·5	1·6	1·9	3·1	3·7	4·7	4·8	6·6	7·5	9·2	
Transport	8·3	8·9	9·4	10·7	12·2	12·5	12·1	13·0	14·4	14·4	
Forestry, Fisheries	1·0	1·2	1·4	1·4	2·0	2·2	2·2	2·3	2·6	3·1	
General Services	26·4	28·3	30·8	33·6	41·1	43·0	44·1	46·4	51·2	58·2	
Post Office	7·8	8·8	9·7	10·1	13·3	13·8	14·9	15·3	17·2	20·0	
Defence	7·1	7·5	8·1	8·7	11·3	11·7	10·4	11·4	12·9	14·4	
Justice	5·7	6·0	6·3	7·3	8·2	8·4	9·3	9·4	10·2	11·8	
Public Service Pensions	5·7	5·9	6·7	7·5	8·2	9·1	9·6	10·3	10·9	12·1	
Other Expenditure	9·6	10·4	10·9	12·2	15·0	16·6	17·1	18·7	20·5	22·8	
Total	139·7	152·3	168·6	186·5	222·0	248·0	270·6	307·0	350·8	411·6	4
Remuneration included in above figures	43·8	45·3	51·2	55·3	71·0	75·5	80·9	84·8	93·9	110·6	1
	1960	1961	1962	1963	1964	1965	1966	1967	1968	1969	
Gross National Product	673	723	778	830	950	1,018	1,071	1,159	1,288	1,444	
Govt. current expenditure as percentage GNP	20·8	21·1	21·7	22·5	23·4	24·4	25·3	26·5	27·2	28·5	

Table 1·1 shows that in every financial year the Minister for Finance planned to balance the current budget. A planned deficit appears for only one financial year, 1968–9, and then in the estimates as revised in the light of the supplementary budget of November 1968, and not in the estimates as presented in April. In all years, with the exception of 1964–5, 1966–7 and 1970–1, there was an allowance of between £2 and £4 million for errors of estimation, expressed as a net deduction from estimated expenditure. With the exception of 1965–6 and 1969–70, the allowance for errors of estimation tended to be made in years in which one of the aims of the current budget was expansion or reflation. In the years when the current budget aimed at curbing inflation, no such allowance was made. In some measure and to a modest extent, the allowances for errors of estimation seem to have been varied so as to work towards a small deficit in years when some reflationary influence was required from the current budget, and towards a small surplus when some disinflationary influence was thought to be necessary.

The increase in annual revenues and expenditures through the period show the effects of automatic and discretionary influences. The private sector and the government current accounts are mutually dependent. Given tax rates and the decisions and norms relating to expenditure, government revenues and expenditures respond automatically to changes in the national income. If incomes rise, so do the receipts from direct taxes; if private expenditures increase, the receipts from indirect taxes will rise; if unemployment falls, current payments to the unemployed will fall; if the sale of subsidised commodities rises, so will government expenditures. These automatic changes provide a measure of the 'built-in flexibility' of the current budget. If current revenues rise faster than current expenditures as the national income rises, and decrease faster than current expenditures as national income falls, then the current budget will act as an automatic stabiliser, moderating the fall in aggregate demand during a recession and curbing its growth during an expansion. In Ireland, the current budget may tend to work as an 'automatic stabiliser' during a recession (e.g. during 1965–6). During an expansion when prices are rising, its net effect may be destabilising. However, in the absence of a model which includes fiscal variables, these conclusions must rest more on hunch than on analysis.

Discretionary changes mean government decisions to change individual items of current expenditure or tax rates. These affect disposable income and expenditure in the private sector, and changes there react back on current revenues and expenditures. It is these discretionary changes in revenues and expenditure which give an indication of actual budgetary policy in each year, and not the historical record of the budgetary outcomes which shows the combined effects of discretionary and automatic changes.

A very crude attempt is made in Table 1·3 to distinguish automatic and discretionary changes in Irish current budgets over the last ten years. For each year, the differences between the actual current revenues and expenditures in the previous financial year, and the estimates of current revenues and expenditures for the financial year ahead as they emerge from the budget statement, are taken as a measure of the combined effect of automatic and discretionary changes. The discretionary changes are equated with the tax and expenditure changes made in the current budget, and the balance of the difference between the outcome for the past year and estimate for the next year is attributed to automatic changes.

TABLE 1.3

Automatic and Discretionary Changes in Current Budgets

(£ million)

| Year | Automatic | | Discretionary | |
	Revenue	Expenditure	Revenue	Expenditure
1960–1	£7·00	£7·85	£0·22	£2·60
1961–2	£5·32	£4·80	£0·59	£2·20
1962–3	£8·14	£8·17	£3·23	£4·49
1963–4	£8·66	£12·25	£9·43	£2·98
1964–5	£23·93	£23·32	£6·78	£5·17
1965–6	£18·22	£17·67	£5·55	£6·02
1966–7 (March budget)	£9·70	£13·93	£12·58	£0·35
1967–8	£20·68	£20·24	£1·60	£6·93
1968–9 (April budget)	£23·05	£26·45	£4·23	£4·62
1969–70	£42·11	£32·89	£5·33	£8·24
1970–1	£51·18	£41·38	£12·70	£21·96

This approach is subject to a number of very obvious defects. For example, some part of the estimated increase in expenditure in each year is due to the introduction of new heads of expenditure based on recent enactments; this increase is not a result of the effects on the budget of changes in the private economy. Indeed, even a cursory perusal of the detailed estimates for any financial year suggests that there may be a fairly wide margin for varying the increases in expenditures within the framework of existing legislation. It seems probable that a sizeable part of at least some increases shown in the estimates is based on 'discretionary' decisions which are influenced by the expected tightness (or otherwise) of the budgetary position for the year ahead. In addition, with this approach increases in public sector wages and salaries which are expected to occur in the year ahead would be regarded as a discretionary expenditure. However, given the policy decision that public sector incomes should be kept in line with those paid for comparable work in the private sector, it could be argued that it would be more appropriate to regard increases in public sector wages and salaries as automatic. The same would apply to some increases in expenditure on health and education. If standards are to be maintained, then increases in the demand for these services (for example, as a result of an increase in the number of children of school age) will require unavoidable increases in the expenditures on them. Again, relatively few 'discretionary' changes have their full impact on current revenues and expenditures in the fiscal year in which they are introduced, and this is especially true of changes in social welfare payments, direct taxation and public sector salaries. This means that relatively small 'discretionary' changes in one year can account for the greater part of the 'automatic' changes in the following fiscal year.

The only argument in favour of the crude approach in Table 1·3 is that it relates discretionary action to what is actually done in the budget after allowance has been made for all automatic changes and all previous 'discretionary' decisions. The discretionary changes in revenue and expenditure never exceeded 5% of total revenue and expenditure, and were generally very much less than this. The figure of 5% was reached for revenue only in 1963–4 and 1966–7—years which followed large current budget deficits.

With the exception of 1963–4 (when corporation profits tax was increased) and 1966–7 (when the standard rate of income tax was raised), the discretionary changes in revenue were made by major increases in indirect taxes (or the introduction of new indirect taxes such as turnover tax and the wholesale tax), often offset in part by minor alleviations in direct taxation. This shift towards indirect taxation was an integral part of government policy throughout the period. The most recent statement of the philosophy lying behind it was given by Mr Haughey in his budget statement in 1970.[12] 'The incidence of direct taxation on individual taxpayers is already relatively high'; moreover, 'income tax is paid by a comparatively small section of the community' and 'can now represent a considerable burden on a wide variety of small incomes.' The system of P.A.Y.E. means that the impact of an increase in income tax 'is little different in its effect on pay claims from a corresponding amount of indirect taxation.' Furthermore, 'direct taxes tend to discourage effort and to have widespread disincentive effects.' If the direct tax base is narrow, it can be extended; indeed, it could be argued that its extension (e.g. to farm incomes) would pose no greater difficulties than those associated with the introduction of turnover tax. If the direct tax burden on small incomes is considerable, so is the burden of indirect taxes, and the latter are on balance regressive. In the absence of empirical tests, it is impossible to judge whether or not the Minister's hypotheses about the incidence and effects of direct (as compared with indirect) taxes are correct.

Table 1·1 shows that the tax and expenditure changes made in each current budget had the effect of bringing the estimates for current revenue and expenditure for the fiscal year ahead into balance. The fact that a balanced current budget was a specific aim of policy was made clear in successive budget statements. Thus, Dr Ryan (*Dáil Debates, 27 April 1960*) conceived it his duty 'to make sure that the current budget is kept in balance.' In 1963, (*Dáil Debates, 23 April 1963*) he stated: 'we must do something effective this year . . . towards bringing the current finances back into balance.' In 1967, Mr Haughey, (*Dáil Debates, 11 April 1967*) stated that 'the government are not prepared to

[12] As early as 1959, Dr Ryan said in his Budget Statement: '. . . the incidence of taxation can be a serious disincentive to individual effort. It can and does militate against managerial, executive and professional ability being fully applied to the raising of the levels of production and employment. It also adds to the comparative attractions offered for such talent outside the country . . . . A lightening of income taxation, sur-tax as well as income tax, is necessary to improve the position.' The Commission on Income Taxation in its third report recommended the introduction of a purchase tax at wholesale level at a rate or rates between $7\frac{1}{2}\%$ and 15% the proceeds to be used to reduce the rate of income tax. See also the *Second Programme for Economic Expansion* (pp. 262–8) and the *Budget Statement 1968* (p. 21) for statements of policy on direct and indirect taxation.

allow this (increase in current expenditure) to destroy the balance achieved in the current budget.' In 1968 (*Dáil Debates*, 23 April 1968), his aim was 'to maintain the balance on current account which has ruled for the past two financial years.' In 1969 (*Dáil Debates*, 7 May 1969) he proposed 'to balance the current account for 1969 at a level which is likely to facilitate the achievement of our economic and social aims.' In years in which there was no specific statement that the aim was to balance the current budget, the estimates of revenue and expenditure emerging from the budget statement were always in balance.

Two main reasons were often given for bringing the current budget into balance or for keeping it in that position. First, the need to contain inflation and especially to reduce (or prevent an increase in) the balance of payments deficit, which was seen as a critical external symptom of domestic inflation. This reason was emphasized by successive Ministers for Finance in each year since 1964, and it tended to get greater emphasis in years which followed deficits in the current budget. The second reason was to husband resources for the public capital programme. Thus, Dr Ryan (*Dáil Debates*, 23 April 1963) stated that the persistence of the deficit on the current budget 'would divert into the financing of current expenditure savings needed to finance the higher capital expenditure which is nationally desirable.' Again, in 1966 Mr Lynch stated (*Dáil Debates*, 9 March 1966): 'The plain fact is that if we run a deficit on current account, the borrowing incurred to finance it will subtract from our capacity to finance capital expenditure.' The general tone of the budget statements throughout the period suggests that this view was important in all years. While these two reasons to contain inflation and to husband resources for the public capital programme were used to justify a balanced current budget, they would together have justified a current budget surplus, especially towards the end of the period.

The aim in the current budget each year was, then, to balance estimated revenues and expenditures. Deficits were unplanned and unintended—they were things that happened during a financial year as a result of unexpectedly low buoyancy in revenue or unexpected increases in expenditure. Once current deficits appeared, action was speedily taken to remove or reduce them—often in the course of the financial year in which they emerged (for example, the supplementary budgets of 1966 and

1968); at the latest, action was taken in the next budget to prevent a deficit in the next financial year. Given the inflationary pressures that characterised most years in the past decade, this policy operated in the right direction. Towards the end of the 1960s, surpluses on the current account might have been desirable. This possibility was aired only in one budget statement. In 1967 (*Dáil Debates,* 27 April 1967) Mr Haughey when discussing the distinction between current and capital expenditures said: 'There is, of course a distinction between the accounting principles involved here and the economic principles which must determine, from year to year, how much of total government expenditure should be financed from taxation and how much from borrowing ... This year, a reflationary rather than a deflationary budget is called for and it would not be appropriate to increase taxation in order to cover any of the items now classified as capital. The question must, however, remain open for consideration in relation to the circumstances of future years.' Even though inflationary pressures grew in strength from 1968 through to 1970, the aim continued to be balance in the current budget. It would be improper, however, to base a judgement of fiscal policy on the posture of the current budget. It is necessary to look also at the capital budgets,[13] since fiscal policy must be

[13] The statutory basis for the distinction between current and capital in the Exchequer Account was section 14 of the Sinking Fund Act 1875. In 1875 the view was taken that all voted expenditure was necessarily 'current' and that the difference between income and expenditure 'above the line' was the true deficit or surplus on the income account of the Exchequer. The practice arose of treating a small amount of expenditure 'above the line' as capital outlay and of deducting it when assessing at Budget time the charge to be made against current revenue. In 1950 a large scale extension of this practice was introduced, with the description 'capital service' (and therefore proper to be met from borrowing) being bestowed rather lavishly. These 'capital services' were authorised by the annual Appropriation Act and appeared 'above the line' in the Exchequer Account. In the Budget of 1952 this equation of capital services with borrowing was abandoned for the reasons (*inter alia*) that (a) it was necessary to reconsider the validity of the description 'capital services' in particular cases, and (b) the extent to which capital expenditure should be met from borrowing could be assessed only in relation to the general economic and financial position. In 1953, the Minister for Finance (Mr MacEntee) questioned the appropriateness of including certain items as capital in the Budget but due to the then existing economic position stated that it was not a year 'for categorically assigning to taxation charges which for some years past have been met by borrowing'.

In the years since 1953, successive Ministers for Finance have probably found themselves in the same position as Mr MacEntee, and voted 'capital services' with marginal additions or subtractions have continued to be financed from borrowing.

judged by reference to total government expenditure and the manner in which it is financed.

The capital budget sets out the capital expenditure of government departments, local authorities and State bodies and the manner in which that expenditure is financed. The details of the capital budget in each financial year from 1960–1 to 1970–1 are set out in Table 1·4, with comparisons between the estimate and the outturn for each year (save the last for which estimates only are given.)

The main reasons why funds were needed are classified under *requirements*. The main requirement was the financing of the public capital programme, which accounted for about 90 per cent or more of the total in each year, except in 1965–6 and 1968–9 when it was around 80 per cent. The second heading on the requirements side in Table 1·4 is provision for debt redemption and the refinancing of borrowings of earlier years by State bodies. This item increased almost ten fold over the period and by 1970–1 consisted mainly of provision for the redemption of maturing government securities. By itself, net debt redemption (i.e. the extent to which holders of maturing securities opt for payment in cash) increases the liquidity of the private (and probably non-bank) sector. It involves the substitution of an exchequer liability to the private sector by one to the banking system or to external lenders, from which the funds needed for the redemption are obtained. The third heading is the deficit on the current budget which reached its highest levels—£8 million and £8·4 million—in 1965–6 and 1968–9 respectively. The final item is the miscellaneous category, the content of which varied from year to year. It includes payments to the World Bank, the International Monetary Fund and the International Development Association, aids to industry (to help meet the cost of the British Import Levy after 1964), government assistance towards the costs of the British Special Import Deposits (after 1968), and net reductions in the public's holdings of Exchequer Bills. This item reached its highest levels in 1965–6 and 1968–9, when the only payments to the I.M.F., I.B.R.D. and I.D.A. were made in this period, and when the largest reductions took place in the public's holding of Exchequer Bills.

The sources of the funds to meet these requirements are shown in Table 1·4 under *resources*. The first heading consists of the

TABLE

Capital Bud

	1960–1		1961–2		1962–3		1963–4	
	Est.	Act.	Est.	Act.	Est.	Act.	Est.	Act
RESOURCES								
Local Authorities and State Bodies Resources of local authorities and State bodies other than Exchequer advances and grants	12·6	10·2	11·5	10·9	15·3	16·4	21·8	26·1
EXCHEQUER Loan repayments	3·6	3·7	3·5	3·8	3·7	5·4	4·4	4·4
Investment resources of departmental funds	8·0	10·6	9·0	13·3	12·0	16·7	14·0	12·5
Small savings and Prize Bonds	12·0	6·4	8·0	7·4	8·5	7·8	8·5	7·0
Miscellaneous	0·5	0·4	0·5	0·5	0·5	0·5	0·5	0·5
Other borrowings: National loans and Exchequer Bills etc.— Public Banks Foreign borrowing	} 19·7	18·7 6·6 — }	24·9	19·7 4·6 — }	27·9	19·1 8·3 }	31·9 }	18·9 14·6
Total	56·3	56·6	57·3	60·1	67·9	74·1	81·0	84·0
REQUIREMENTS Expenditure on public capital programme	54·4	50·6	55·5	57·2	66·9	65·1	79·7	78·5
Refinancing of borrowings of previous years	1·7	1·8	1·3	1·5	0·6	1·6	0·8	0·8
Current budget deficit	—	0·73	—	0·71	—	4·85	—	2·22
Miscellaneous	0·3	3·5	0·5	0·6	0·4	2·5	0·6	2·5
TOTAL	56·3	56·6	57·3	60·1	67·9	74·1	81·0	84·0

*Estimate only.

†This estimate was cut by £3·4 million in October, 1965.

t to 1970–1 (£ million)

	964–5		1965–6		1966–7		1967–8		1968–9		1969–70		1970–1*
	Act.	Est.	Act.	Est.	Act.	Est.	Act.	Est.	Act.	Est.	Act.	Est.
	29·7	31·4	28·1	30·5	33·0	32·1	35·4	46·2	52·6	66·0	63·8	73·8
	4·8	5·7	5·8	6·3	6·2	6·7	5·7	6·2	11·0	7·3	7·7	8·4
	16·4	17·0	12·9	14·5	14·3	19·0	20·3	23·0	17·3	25·6	28·0	30·0
	9·7	10·0	2·3	7·0	7·0	9·0	6·2	7·5	5·0	6·0	2·9	6·0
	0·6	4·35	3·7	0·2	1·2	2·0	2·0	0·3	16·0	2·8	3·0	5·4
	23·0 ⎫ 19·0 ⎬ ⎭	43·6	15·3 ⎫ 33·3 ⎬ 14·9 ⎭	22·0 28·0	24·8 ⎫ 16·2 ⎬ 9·8 ⎭	24·0 25·0	24·7 25·0 —	24·0 33·0 1·0	25·5 ⎫ 52·0 ⎬ 2·8 ⎭	26·0 53·0	19·2 ⎫ 50·0 ⎬ 10·9 ⎭	25·0 75·0
	103·3	112·1	116·3	108·5	112·5	117·8	119·3	141·2	182·2	186·7	185·5	223·6
	97·8	103·7†	99·3	97·8	99·0	108·5	111·4	136·4	141·6	166·8	173·4	194·5
	0·5	2·1	3·4	7·4	8·3	7·1	6·6	2·8	10·7	11·9	8·9	16·2
	4·07	—	7·8	—	—	—	—	—	9·0	—	0·5 ⎫ ⎬ 2·7 ⎭	12·9
	0·9	6·3	5·9	3·3	5·2	2·2	1·3	2·0	20·9	8·0		
	103·3	112·1	116·3	108·5	112·5	117·8	119·3	141·2	182·2	186·7	185·5	223·6

internal resources of local authorities and State bodies (for example, from annual depreciation allowances or increases in pension funds) and monies raised by them from banks, insurance companies or stock issues. These latter might be raised within Ireland by borrowings from the non-bank public (and these were probably the main sources in the first half of the period), or from Irish banks[14] and external borrowing (and this was increasingly the case in the latter half of the period), and these borrowings were guaranteed by the State. These resources provided by local authorities and State bodies provided about one-third of the total expenditure on the public capital programme from 1963-4 onwards. The second item—loan repayments—is self explanatory. The third item is the investment resources of departmental funds. This consists of that part of the annual contributions to sinking funds (which is a Central Fund charge on the current budget) that is not applied to the purchase (i.e. cancellation) of the government securities to which the funds relate; of the net interest on the Post Office Savings Bank Fund, the Social Insurance Fund and other departmental funds, and net sales of government securities held in Departmental Funds. Variations between the estimate and outcome of the contribution to the capital budget from Departmental Funds is probably explained by unexpected movements in the last of these three constituents—for example, net purchases (sales) of government securities rather than the net sales (purchases) which were anticipated when the capital budget was being prepared. The fourth item—small savings and prize bonds—consists of the net increase in deposits with the Post Office Savings Bank and the Trustee Savings Banks, and sales less redemptions of prize bonds, during each year. The fifth is a miscellaneous category: the small sums available under this head up to 1964-5 came from the National Development Fund; the entries for 1965-6 and 1968-9 related to borrowing from the Central Bank to meet payment to international bodies (such as the I.M.F.) and were offset by corresponding items on the requirements side of the account; the figures for the remaining years represent miscellaneous minor borrowings.

The final item—other borrowings by the Exchequer—shows the proceeds of the sale of national loans to the public and of

---

[14] In the published statistics for Associated Bank lending, local authorities and State bodies are included in the private sector.

borrowing from the Irish banking system and from abroad. During this period the annual sales of new government securities to the public lay in the range £18½ to £25½ million. Exchequer borrowing from the banking system rose from less than £5 million in 1961–2, to over £33 million (which included £20 million from the Central Bank) in 1965–6; it fell to just over £16 million in 1966–7 and rose rapidly to over £50 million in 1968–9 at which level it remained in 1969–70. Foreign borrowing by the Exchequer first appeared in 1965–6, providing more than £15 million. It was also important in 1966–7 and 1969–70 at around £10 million. The counterpart in Irish currency of foreign borrowings, including the drawings of £8 million from the I.M.F. in 1965–6, were used to finance public expenditure.

In Table 1·5, the sources of finance for the capital budget (as set out in Table 1·4) are summarised under three main heads:

TABLE 1.5.

Summary: Capital Budget Resources

	1960–1	1961–2	1962–3	1963–4	1964–5	1965–6	1966–7	1967–8	1968–9	1969–70	1970–1*
Resources						£ million					
"Internal"†	24·9	28·4	38·9	43·5	51·5	50·5	54·7	63·4	96·9	102·5	117·6
Borrowing from non-bank Irish public	25·1	27·1	26·9	25·9	32·7	17·6	31·8	31·0	30·5	22·1	31·0
Borrowing from the banking system and abroad	6·6	4·6	8·3	14·6	19·0	48·2	26·0	25·0	54·8	60·9	75·0
Total	56·6	60·1	74·1	84·0	103·2	116·3	112·5	119·4	182·2	185·5	223·6

*Estimate.
†Internal resources include resources of local authorities other than Exchequer advances and grants, on repayments, investment resources of Departmental Funds and the miscellaneous item in Table 1.4.

'internal' resources (i.e., resources of local authorities and State bodies other than Exchequer advances and grants, loan repayments, investment resources of Departmental Funds and the miscellaneous item), borrowing from the non-bank public (i.e. small savings, prize bonds and the sales of national loans and Exchequer Bills to the public), and borrowing from the banking system and from abroad. The first of these items is overstated (and the second and third correspondingly understated), by amounts rising from just less than £5 million in 1960-1 to over £35 million in 1969-70 and nearly £46 million in the estimate from 1970-1, by the inclusion in it of State-guaranteed borrowings of local authorities and State bodies from the public, from banks and from abroad, and of miscellaneous minor borrowing from undivulged sources. It is also inflated by the inclusion in it of payments to the I.M.F. of £4·35 million in 1965-6 and £16 million in 1968-9: these payments were matched by corresponding payments in foreign currencies on the requirements side of the capital budget. The broad picture which emerges is an increase of about fivefold in 'internal' resources and little change in the amounts borrowed from the non-bank public, with borrowing from the banking system and from abroad acting as the residual source of funds and increasing tenfold.

Four main conclusions may be drawn from Tables 1·4 and 1·5 and from the relevant budget statements. *First,* the capital budget was regarded as the principal instrument for achieving the faster growth which was the main aim of government economic policy throughout the period. The role attributed to the capital programme was not without justification. By 1970-1 about 35–40 per cent of the public capital programme—advances and grants for private housing, industrial and agricultural grants and loans—went directly to finance capital expenditure by the private sector. In addition, the availability of these grants and loans induced substantial investment from private sector funds.

In this current financial year, the public capital programme is estimated to be directly responsible for more than half of gross domestic fixed capital formation.[15] The importance of the capital budget for economic growth was emphasised in the annual budget statements throughout the 1960s. Thus, Dr Ryan (*Dáil Debates,* 19 April 1961) argued that 'intelligent investment, at the highest sustainable level, is a necessary condition for sound and rapid national progress.' Again, in 1963 (*Dáil Debates,* 23 April 1963) he stated: 'Clearly it is right that the State should, directly and indirectly, increase the volume of productive investment and do everything else in its power to promote as rapid a growth of the economy as can be sustained without excessive strain on the balance of payments.'

[15] See Capital Budget 1970, Part II, published in *Budget 1970.*

Second, Ministerial statements taken at their face value seem to suggest that the size of the annual public capital programme was not determined by reference to the expected balance between aggregate demand and supply in the year ahead, but rather to an estimate of the funds that could be obtained to finance the work that it was felt had to be done. For example, 'This year's programme is heavier but with the support of the community which, I am sure, will be generously given, I expect that we shall be able, no less successfully, to raise, from our own resources, the finance necessary to ensure its completion.' (Dr Ryan *Dáil Debates,* 27 April 1960). Again, 'A substantial volume of public investment is planned for 1961–62 but not, I believe, more than national savings can be relied upon to finance.' (Dr Ryan, *Dáil Debates,* 19 April 1961). In 1967 (*Dáil Debates,* 27 April 1967) Mr Haughey explained that: 'The decision to expand the public capital programme is based on the expectation that normal home resources will continue to increase and that resources will not have to be drawn upon to finance any sizeable deficit on current account.' In 1970 (*Dáil Debates,* 22 April 1970) the Taoiseach stated: 'The public capital programme dominates national capital formation. It is therefore necessary to provide it with sufficient funds to maintain and, if possible, to increase national investment.' During the 1960s, the emphasis in Ministerial statements seemed to change: in the earlier years, the emphasis was more on constraining public capital expenditure within the limits of national savings; towards the end, as the words of the Taoiseach (deputising for Mr Haughey) in 1970 show, the emphasis was rather on the necessity of providing the public capital programme with sufficient funds to maintain or increase it.

However, it would be fairer to recognise that these Ministerial statements were perhaps more in the nature of exhortations than descriptions of a fiscal philosophy. A closer approach to this latter is to be found in the White Paper on Public Capital Expenditure published in October 1965. It was there stated that by the mid-1960s 'the earlier problem of a surplus of capital in relation to development ideas was rapidly transformed into a problem of too many development projects in relation to the capital immediately available . . . indeed, the forces generating demand and causing increased spending from incomes and credit grew so strong that resources—current production and capital inflow—became insufficient to match them. The possibility of demand becoming excessive was a risk . . . deliberately taken by the government in the conviction that all available resources should be used to promote the fastest possible rate of progress in output and employment.' This policy of 'running on the outside of the fiscal track' is a valid one for a developing economy. However, the White Paper went on to state that only experience could establish 'what level of resources . . . could be continuously counted upon . . . and how fast could money incomes, credit and capital expenditure expand without creating excessive pressure on resources.' By 1969–70, and certainly by 1970–1, experience had already established that the limits for domestic demand pressures and for domestic incomes inflation had both been passed. Nevertheless, the posture of fiscal policy remained unaltered.

If the quotations in the paragraph before last seem to suggest that public capital expenditure was not assessed in the context of the expected balance between aggregate demand and supply, there are others which show that it was in fact so weighed. For example, as early as 1953, Mr MacEntee stated (*Dáil Debates*, 6 May 1953, Column 1202) that 'the contribution by the State towards the stimulation of economic activity is determined, not by deficits on the current budget but by the overall effect of State expenditure capital as well as current.' Similar statements were made by Mr Lynch in 1965 and 1966, and by Mr Haughey in 1967, 1968 and 1969. Together, these would be consistent with the following fiscal philosophy. The current budget was not seen as an effectively available instrument of demand management. Current expenditures were inevitably on a rising trend; once they had risen, they could not be reduced because of their predominantly social and redistributive character. The only effective way of arresting their growth was by insisting on increases in current expenditures being fully met from increases in tax revenues. Demand management could best be done by varying the rate of growth in public capital expenditures. This was a much better way of stimulating demand, or dampening its growth, than running current deficits and surpluses. Despite the difficulties associated with using capital expenditures as an economic regulator, they were in fact used in this way on several occasions—for example, in 1965–6 and 1966–7 and to a much smaller degree in 1969–70.

The philosophy set out in the preceding paragraph is consistent with Ministerial statements. Within it, the emphasis seemed to be more on variations in public capital expenditures rather than on the net effects on aggregate demand of these variations together with the way in which the capital expenditures were financed. For example, as a disinflationary measure, the growth in the public capital programme was reduced from 25 per cent in 1964–5 to 1½ per cent in 1965–6. Between these two years, however, the residual borrowing from the banking system and from abroad rose from £19 million to over £48 million, and this must have gone some way towards offsetting (if not exceeding) the disinflationary effects of the slower growth in expenditures. Again, in 1968–9 when it might have been prudent to take early action to dampen the growth in aggregate demand, the public capital programme was increased by almost 28 per cent (or £30 million). Between 1967–8 and 1968–9, however, the residual borrowing rose from £25 million to nearly £55 million. Finally in 1970–1 the growth in the public capital programme was reduced to just over 12 per cent as compared with 23 per cent in 1969–70, and the anticipated residual borrowing rose by over £14 million to £75 million. Even allowing for the supplementary budget of 28 November 1970, the residual borrowing could exceed this figure in the current financial year, so that the net impact of the public finances on the economy may be expansionary, and not disinflationary as seemed to be the intention.

Third, borrowing from the banks and from abroad became progressively more important during the period as a source of funds. It accounted for over 11 per cent in 1960–1, nearly 8 per cent in 1961–2 and an estimated one-third in 1970–1, of the total. If requirements rose faster than expected, or if the resources becoming available under other heads fell short of the estimates, the deficiency was made good by increased borrowing from the banks and from abroad. Given the growth in the public capital programme, the growth in this source of funds is in large part explained by the relative stability in the funds raised by borrowing from the non-bank public (see Table 1·5). Between 1960 and 1969, there was a threefold increase in net savings (i.e. excluding depreciation provisions), yet over this period the sums raised by borrowing from the non-bank public fell by £3 million from £25 million in 1960–1 to £22 million in 1969–70. This suggests that the terms offered by the government became progressively less attractive, as compared with the terms offered by other borrowers, through the 1960s.

The sluggishness in the yield from borrowing from the non-bank public was used to justify greater reliance on bank borrowing. Throughout the period and especially during its latter years, time deposits held in Associated Banks increased, and at least some part of this increase was regarded as a 'diversion' of the growth in national savings away from small savings and national loans. Therefore, 'a contribution by the commercial banks to the financing of public capital expenditure is legitimate, because in Ireland these banks receive a large part of the current increase in public savings.' (Mr Lynch, *Dáil Debates*, 9 March 1966). It is difficult to assess the substance of this line of argument. If the 'diversion' of deposits from (for example) the Post Office Savings Bank to the commercial banks had not occurred, it is probable that *total* deposits in the latter would have risen by much the same amount.[16] The proceeds of the increased deposits in the P.O.S.B. would have been immediately available to the government and would have been reflected in increases in the government's current account with the Bank of Ireland. The effects of the expenditure of this new money on total commercial bank deposits held by the private sector would have been much the same, irrespective of whether the deposit had been made in the first instance with the P.O.S.B. or with a commercial bank. It is possible, however, that the balance between current and deposit accounts would have been different as between these two positions, but it is very difficult to demonstrate how great the difference would be. Moreover, the official line of argument was both partial (necessarily so at the time)[17] and deficient even on its own plane: partial in that it looked only to increases in deposits with the Associated Banks; deficient in that it seemed to regard the whole of the rise in time deposits as coming from current savings, ignoring the extent to which it may have been due to a shift in private portfolios towards more liquid assets.

[16] See: National Board for Prices and Incomes: Report No 34, *Bank Charges,* HMSO, Cmnd. 3292.

[17] Figures for deposits, etc., in non-Associated banks were first collected by the Central Bank for December 1966, and were first published in July 1967.

The argument described in the previous paragraph can be presented in more general terms. The commercial banks in Ireland were the custodians of the nation's savings to a much greater extent than in other countries. By borrowing from the banks (e.g., to finance grants and loans), the government channels savings to the private sector for desirable investment—a function that would be performed elsewhere by the Stock Exchange or other intermediary. However, this argument tends to equate the growth in the resources of the commercial banks with increased current savings, and ignores the effect which government borrowing from these banks may have on the money supply. More important, this argument—like that in the previous paragraph—by-passes what may have been the real problem, namely that the terms offered by the government to the non-bank public became progressively less attractive than those offered by other borrowers. This provides the most satisfactory explanation of the relative stability in the funds obtained from this source during the past ten years. The problem might have been solved if the government had offered the non-bank public the same terms as those on which it borrowed from the Associated Banks, thus enabling the public to enjoy the difference between the banks' borrowing and lending rates.

The notion that it was legitimate for the government to expect a contribution towards the public capital programme from the commercial banks was usually associated with the notion that the new resources of the banks should be shared fairly with the private sector (in which local authorities and State bodies were included). Thus, in 1966, Mr Lynch (*Dáil Debates,* 9 March 1966) stated that he 'would not seek to defend a situation in which the new resources of the banks were claimed almost exclusively by the public sector and credit had to be unduly curtailed for productive private needs.' In large part, this may be a non-problem, because funds borrowed by the government are very quickly channelled back into the private sector. When concern was expressed, the real problem was not so much the distribution of any increase in bank credit between the Exchequer and the private sector as the fact that the total demand for new bank credit exceeded the increase which the Central Bank for the time being deemed permissible. The government was in a position to make its demands effective. At the same time, the Associated

Banks, operating in an increasingly competitive environment, were loathe to refuse accommodation to their clients. The conflict was usually resolved by an increase in the amount of new credit extended beyond the Central Bank's limit, sufficient to accommodate both the government's needs and the banks' desire to maintain contact with their customers. These issues arose in their most acute form in 1965–6. In that year, it was estimated that £20 million would be required from the banking system and from foreign borrowing. In the event, because funds from other sources fell far short of the estimates and requirements rose significantly, £46½ million was required from the banks and external borrowing. This was obtained from the Central Bank (£20 million), by making a drawing from the I.M.F. to support national reserves (£8 million), by the issue of a Sterling/Deutschemark Loan (£7 million) and by borrowing from the Associated Banks (about £11 million). The exceptional finance from the Central Bank enabled the government's requirements to be met without bank credit for the private sector being unduly curtailed.[18]

The *fourth* conclusion that may be drawn is that the capital budgets may on balance have been destabilising. There was a tendency to press public capital expenditure up to the limit of

[18] The Annual Report of the Central Bank for 1965–6 gives a more detailed account on pages 45–6 of what happened: 'At the beginning of September 1965, the Minister for Finance indicated additional borrowing requirements of £18·0 million from the banks. To meet these requirements in full, while complying with the Central Bank's advice on credit policy, would have involved severe curtailment of credit to the private sector. In these circumstances an arrangement was made under which the Central Bank, as an exceptional measure, purchased on 30 September from the Associated Banks £20·5 million (nominal) 6% Funding Loan, 1969, which was issued by the Minister for Finance for this purpose. Immediately prior to this operation the total amount of outstanding Exchequer Bills which had been issued directly to the Associated Banks was £40·3 million, of which £13·7 million had been rediscounted and was held by the Central Bank. Following the issue of the Funding Loan the Minister paid off £20·0 million Exchequer Bills which matured on 30 September, namely the £13·7 million held by the Central Bank and £6·3 million held by the Associated Banks. On the same date the Associated Banks took up £18·0 million fresh Exchequer Bills. As a result of these transactions there was an increase of £11·7 million (£18 million *less* £6·3 million) in the Associated Bank's holding of Exchequer Bills.'

In addition to the £20 million provided in this way, the Minister for Finance obtained another £8 million from the Central Bank as a result of the transactions associated with the drawing from the I.M.F. (The latest figures show borrowing from the banking system and from abroad somewhat higher at over £48 million in 1965–6).

the funds that it was thought would be available or obtainable, on terms that the government was prepared to contemplate. Together with the expansionary effects of the growing (though in intent balanced) current budget, this ensured that the level of aggregate demand was generally buoyant and often excessive. This contributed in its turn towards rising deficits in the current balance of payments (which were acceptable for as long as they were financed by voluntary net capital inflows) and rising prices and money incomes (which required substantial increases in public capital expenditure if it was even to be maintained in real terms). In these circumstances, the economy was vulnerable if capital inflows fell or growth slackened, and this was shown in 1965–6. Even though the increase in the public capital programme at current prices was cut to $1\frac{1}{2}$ per cent (as compared with 25 per cent in 1964–5), there were great difficulties in financing it. Residual borrowing by the Exchequer rose to almost £47 million as compared with £19 million in 1964–5; and while there was little change in the 'internal' resources of local authorities and State bodies, the extent to which these were obtained from external borrowing rose significantly. The disinflationary intention of the smaller increase in the capital programme may therefore have been offset by changes in the manner in which it was financed. In the following year, 1966–7, public capital expenditure was reduced by about 1·7 per cent as compared with 1965–6 (implying a fall of a larger order in real terms) and a much closer correspondence was achieved between estimate and outcome on both the resources and requirements sides of the capital budget. However, the residual borrowing from the banking system and from abroad was reduced to £26 million, so that the disinflationary impact may have been greater than was intended. Thereafter, an expansion developed which was in many respects similar to that which culminated in 1964–5. The sequence of expansion and an accelerating rate of increase in prices, followed by correction which was to some extent at least enforced by financial exigencies, probably made for a slower average rate of growth in real GNP. This sequence made for a somewhat higher average annual increase in prices and money incomes, because the accelerating increases in prices and in the rates of contractual incomes during the expansion were not offset by any offsetting reductions when corrective measures were being applied. If the

management of aggregate demand had been more effective
during the expansion phase, there would have been less need for
correction.

SUMMARY AND CONCLUSIONS

The use of fiscal policy for demand management requires that
changes in the government's current and capital expenditures and
in its revenues (from taxes, borrowing from the non-bank public
and borrowing from the banks and abroad) are determined by
reference to their impact on the level of aggregate demand. In
the most general terms, if inflationary pressures are expected to
develop, the growth in total government expenditures should
be moderated and a larger proportion of the funds required to
finance them raised from taxation and by borrowing from the
non-bank public; if it is thought that aggregate demand will
become deficient then (assuming the balance of payments is
acceptable) the growth in public expenditure should be acceler-
ated and/or a smaller proportion of them financed by taxation
and borrowing from the non-bank public. If fiscal policy is
used in this way, in conjunction with monetary and other policies
that are consistent with it, the average annual rate of growth in
prices, money incomes and external deficits will be moderated
in periods of expansion, and taking periods of expansion and
recession together the average annual rate of price—and income—
inflation will be lower and the real growth rate higher. Conflicts
between the aims of demand management and the objectives of
growth and full utilisation of productive resources will arise only
when price—and income—inflation occur when the economy is
operating at less than capacity. A choice between fuller employment and price-stability must then be made: if the latter is
chosen, proper demand management can contribute towards its
achievement; if the former is chosen, other policies will be
required to contain the balance of payments deficits.

In Ireland in the 1960s, the Budget Statements of successive
Ministers for Finance do not always suggest that the role of fiscal
policy was thought of in these terms. Demand management,
however, was frequently mentioned as an aim of the current
budget, but always strictly within the context of a balanced
current budget. Demand management within the framework of

a balanced current budget requires a delicate appreciation of the effects of different taxes and expenditures which was not shown in the budget statements. Without exception, the aim was to balance the budget in each fiscal year. Action was usually quickly taken to arrest any tendency towards the appearance of a current budget deficit, but the emphasis was often less on the impact of a current deficit on aggregate demand than on the fact that it would usurp a part of the funds available to finance public capital expenditure. Public capital expenditure was regarded as the main cause of economic growth and seemed to be mainly determined by reference to the funds that were expected to be available (or obtainable) to finance it. When it was feared that demand pressures were excessive (as for 1970–1), only the impact of the expenditure side of the capital budget seemed to be considered relevant, and no reference was made to the effect on aggregate demand of the extent to which it was planned to raise funds from borrowing from the banks and from abroad.

The difference between how fiscal policy should have been used and how it was in fact applied is not to be explained by any lack of awareness of the former in the minds of those responsible for the latter.[19] The explanation must lie in the reasons why the balanced current budget rule was followed, in the full knowledge that taken in any one year by itself balance (as opposed to deficit or surplus) in the current budget could have been economically injurious. The explanation must also lie in the reasons why the impact on aggregate demand of the manner in which the capital budget was financed were generally ignored.

The balanced budget rule may have been followed for a number of reasons. *First,* it may have been regarded as good national housekeeping. The good housewife keeps her expenditures within the limits of her housekeeping allowance. If she does not, she runs into debt and the dire consequences for herself and her family will ultimately come home to roost. This analogy between

---

[19] See, for example, *Second Programme for Economic Expansion,* Part II, (Pr. 7670). 'It is desirable that total community spending—capital and consumer spending in due proportions—should be sufficient to keep productive capacity at the highest rate consistent with a competitive level of prices and costs and a reasonable degree of external balance,' (page 295, paragraph 1). '. . . it may be necessary to depart, at the margin, from the traditional methods of financing in order to give greater flexibility in the means available for moderating tendencies towards inflation or deflation,' (page 273, paragraph 8).

national and private housekeeping would have been recognised as an example of the fallacy of composition by those responsible for budgetary policy. However, the public at large, including many who make their living by lending to households to enable them to spend in excess of their current incomes, would probably consider this sort of analogy as having some relevance. Public opinion may therefore have acted in some sense as a constraint on budgetary policy.

Second, and more important, there may have been a feeling that once the balanced budget rule was relaxed, all discipline on the public finances would disappear, and rising deficits and accelerating inflation would follow. One part of this argument would be fallacious, namely, that current budget deficits are always and necessarily inflationary. Inflation would follow only where the economy's resources were already fully utilised, and not where idle capacity existed (except where this was itself associated with continuing price—and income—inflation). Its main assertion, however, rests on an assumption about the nature of a democratic government in a modern western economy. If there were no balanced budget rule, there would be no limit on government expenditure. But if the rule is strictly applied, the political advantages of more spending have to be balanced against the political disadvantages of the additional taxation needed to finance it. In other words, governments will only then continue spending 'until the marginal vote gain from expenditure equals the marginal vote loss from financing.'[20] The description of the role of fiscal policy in demand management in Section II above contains no simple and easily comprehensible formula for containing government expenditures similar to the balanced budget rule.[21] If the latter goes it would be argued that probity and financial discipline may go with it. However, if the first deviation from the balanced budget rule were to introduce a surplus, this would have met the argument against 'irresponsible' government.

In fact, adherence to the balanced budget rule, taken by itself, generally worked in the right direction during the 1960s. In all

[20] Anthony Downs, *An Economic Theory of Democracy*, New York, 1957, 73.

[21] It is worth noting that the balanced budget rule does not prevent *wasteful* government expenditures—it merely ensures that if they occur, because of pressures from interest-groups in the public or private sector, they will be paid for from taxation.

years with the exception of 1966-7, deficits would have been inappropriate, and the action taken either to prevent deficits from appearing, or to remove them once they had appeared, made inflationary pressures less than they otherwise would have been. However, the balanced budget-rule was apparently equally successful in preventing the surpluses on the current budget, which might have been appropriate in 1964-5 and would certainly have been desirable in 1968-9 and the years since then. This is less easy to understand, since the arguments used against a deficit often seem to make a surplus even more desirable than balance in the current budget.

There was no discipline similar to that provided by the balanced budget convention to govern the capital budget. It would be difficult to conclude from Ministerial statements and what was actually done that the requirements of demand management influenced the successive capital budgets: the aim seemed to be to keep public capital expenditure at as high a level as could be financed. In a developing economy, it is difficult not to sympathise with the desire to keep investment (both public and private) high and buoyant. But this does not in itself fully justify the level of public capital expenditure, because even its economic components were not all of a productive character, promising a yield commensurate with the interest payable on the borrowed funds with which it was financed. The fault lay also with the manner in which public capital expenditure was financed. The extent to which the capital programme was financed from taxation was limited to that part of the annual accrual to sinking funds (a charge against current tax revenue) which was made available for this purpose through the Departmental Funds, and probably provided no more than 5 per cent to 7 per cent of the annual requirements throughout the 1960s.[22] The proportion raised by Exchequer borrowing from the non-bank public fell from about 44 per cent in 1960-1 to just under 11 per cent in 1969-70 and a projected 13 per cent for 1970-1. The proportion raised by Exchequer borrowing from the banking system and from abroad rose from 11 per cent in 1960-1 (and under 8 per cent in 1961-2) to 33 per cent in 1969-70, with a similar percent-

---

[22] The percentages for non-bank borrowing and for borrowing from banks and from abroad understate public sector borrowings because borrowings by local authorities and State bodies are not included. See above.

age contribution projected for 1970–1.[23] This shift in emphasis from borrowing from the non-bank public towards borrowing from the banking system and from abroad added to the demand pressures in the economy, because no significant offsets in private demand were associated with the latter. In other words, the rising expenditures on investment were not offset by reductions in expenditures on consumption, so that aggregate expenditure rose excessively.

In the latter half of the period, fiscal policy was often not consistent with monetary policy. When the demands of the public sector could not be met within the credit ceiling without restricting the private sector, (as in 1965–6 and since 1968), the public sector sought external funds, and as these were spent, the domestic credit supply increased beyond the limits of the Central Bank's advice. When fiscal policy is expansionary, monetary policy cannot for long succeed in being restrictive if the funds to finance the expenditure commitments of the Exchequer are to be provided.

These conclusions about the use of fiscal policy in Ireland over the last ten years may seem harsh, but a recent O.E.C.D. study[24] has shown that not dissimilar strictures would apply to the use of fiscal policy in a number of western countries over the same period. The full recommendations of that study, which are too lengthy to summarise here, are equally applicable to Ireland.[25] A first step is to relax rigid fiscal rules (such as that the current budget should always be balanced), because 'a government's ultimate responsibility is to balance the economy as a whole rather than to balance its own accounts; and rigid fiscal rules are not only insufficient but are a hindrance to this end.' To ensure that policy 'is based on economic realities and needs rather than on political expediency . . . an informed public opinion is essential and governments should play an active role in nurturing it.'[26]

[23] The narrow direct tax base and the price-raising effects of indirect taxes will of course, impose some limit to the contribution from current taxation towards capital expenditure, but this limit will not be as low as the 5 per cent–7 per cent that characterised the 1960s.

[24] Walter Heller et alia: *Fiscal Policy for a Balanced Economy: Experience, Problems and Prospects* O.E.C.D. Paris, 1968.

[25] *Ibid.,* VIII.

[26] The quotations in this and the next paragraph are from Heller et al., *op.cit.* 146, 158 and 159.

To this end, 'a correct appraisal of current economic trends' is needed. Economic forecasts should be published in greater detail and with fuller documentation than is now the practice. These forecasts should (*inter alia*) attempt, on the basis of current policies, to identify the gap between aggregate demand and aggregate supply in the period ahead (see page 7). The forecasts would provide the framework within which the annual current and capital budgets could be judged. The annual budget statement should include an analysis of the impact of current and capital budgets of the previous year on the economy, and explicit estimates of the impact on the economy of the current and capital budgets proposed for the year ahead.

However, it is not necessary to spell out what the use of fiscal policy for demand management would require: the need for demand management, the contribution that fiscal policy can make towards it, the improvements in statistical information and the changes in administrative arrangements that would be necessary to facilitate it, have all been set out in various official publications, notably, the *Second Programme for Economic Expansion*[27] and the *Third Programme for Economic and Social Development*.[28] What is needed at this stage is a fuller awareness in the community at large of the importance of managing demand. If this is not achieved, the 'political' costs of demand management to control inflation may continue to be allowed to outweigh the immediate and prospective economic benefits that would follow from it.

[27] Government Publications Sale Office, Dublin 1964 (Pr. 7670).
[28] Government Publications Sale Office, Dublin 1969 (Prl. 431).

CHAPTER II

Some Issues of Irish Debt Policy

ALAN A. TAIT

THE national debt exists in a peculiar policy limbo. It is neither fish nor fowl. For the government it is a way of getting control over real resources without taxation. To the Central Bank it can be extension of monetary policy and changes in the national debt affect interest rates and all forms of lending. To the people of the country it is a mystery. It is a debt and that in itself must be bad; it is national debt and therefore affects everyone and that makes it worse. Everyone views the national debt in a different way and this can lead to differences of opinion over what constitutes proper debt policy.

One suggestion of this chapter is to look at debt policy as a bridge between pure monetary and pure fiscal policy. Irish debt policy is discussed in the framework normally used for discussion of fiscal policy—in terms of allocation effects, the effects on the distribution of income and wealth, and its role in stabilisation policy. In each of these categories debt policy carries monetary policy implications, but it is hoped this presentation puts the emphasis on debt as a complement to fiscal policy rather than as a pure adjunct of monetary policy. It also puts some problems of the Irish national debt in a context in which they are not usually discussed. Before these considerations there must be a description of the size and structure of the Irish national debt to put the rest of the discussion in perspective.

THE SIZE AND STRUCTURE OF THE IRISH NATIONAL DEBT

Table 2·1 shows the maturity of the Irish national debt from 1923–24 to 1969–70. 1923–24 is the first year in which there is an identifiable national debt. Thereafter it is given for ten year intervals up to 1968–69 when the last two years are shown. For more recent years (from 1949–50) there are three totals given for the size of the debt. The line (a) represents the official figure for debt, and these are the figures to which most references are made. However, it was thought interesting to include some other securities issued by government agencies, and backed by the government, which could be considered as part of the national debt. These are shown in Table 2·1 line (b). There are no such figures for the years before 1949–50, but thereafter they become quite substantial. The third line (c) is simply the total of lines (a) and (b) and represents the largest figure for debt backed by the government.

TABLE 2·1

Maturity of the Irish National Debt 1923–24—1968–70★

£ million

Year	Floating Debt	Marketable Securities (years to redemption)				Savings	External	Total
		−5	5−15	15+	Undated			
1923–24 (a)	2·8	—	—	10·0	—	0·9	—	13·7
1929–30 (a)	3·2	—	8·4	6·9	—	5·8	—	24·3
1939–40 (a)	—	—	—	40·4	—	7·7	—	48·1
1949–50 (a)	15·3	—	27·6	39·9	—	16·9	21·8	121·5
(b)	7·7	—	11·7	55·5	—	0·3	—	75·2
(c)	23·0	—	39·3	95·4	—	17·2	21·8	196·7
1959–60 (a)	116·6	15·3	105·3	75·6	—	42·4	39·5	394·7
(b)	6·2	1·0	17·2	49·0	2·0	1·5	—	76·9
(c)	122·8	16·3	122·5	124·6	2·0	43·9	39·5	471·6
1968–69 (a)	110·3	156·4	124·5	379·5	—	88·5	54·8	914·0
(b)	18·1	—	27·5	75·2	20·5	0·6	14·5	156·4
(c)	128·4	156·4	152·0	454·7	20·5	89·1	69·3	1070·4
1969–70 (a)	111·5	211·6	129·9	379·2	—	107·1	69·6	1008·9
(b)	4·6	16·8	17·2	50·3	55·6	2·5	35·9	182·9
(c)	116·1	228·4	147·1	429·4	55·6	109·6	105·5	1191·8

★ (a) official figures; (b) possible additions; (c) alternative totals—(see text).

For each year the debt is divided into four major parts; floating debt which consists of Exchequer Notes (Treasury Bills) and Ways and Means advances, then Marketable Securities are shown for four categories, under 5 years, 5 to 15 years, over 15 years and undated. Savings are shown as a separate category because for policy making they are different from other marketable securities in that their issuance and redemption are not wholly within the control of the government; for example, holders of national savings certificates can encash them at any time (although they are encouraged by penalties for prior encashment to hold them to full maturity). Finally a column is shown for external debt which, as subsequent discussion will amplify, has important policy implications for Ireland.

The figures in line (b) represent guaranteed loans backed by the government which might be added to the official figure for the national debt; these are mainly land bonds and loans to semi-state bodies. Land bonds could be viewed as a type of Prize Bond because interest is given by redemption on 'drawings' on a lottery basis which can extinguish the bond in anything between 45 to 66 years. However, the State has control over the amount of debt issued (unlike Prize Bonds or small savings) and so the land bonds are treated as a stock having a life of over 15 years. The State also guarantees some semi-state bodies' overdrafts and these were treated as a continuously rolling short term floating debt.

The only undated stock appears in line (b) for the years 1959–60, 1968–69, and 1969–70, and these are loans raised by Bord na Mona 'for its statutory purposes' and as no redemption date is given the category of undated stock was used, although such loans are obviously not marketable.

From Table 2·1 it can be seen that the national debt has risen from £13·7 million in 1923–24 to an official total of £1,008·9 million in 1969–70. If the guaranteed loans are added the figure for 1969–70 rises by £182·9 million to £1,191·8 million of government backed debt.

Table 2·2 shows the maturity of the Irish national debt 1923–24 to 1969–70 in terms of the percentage each type of stock represents of the total. In the early years of the State the national debt had little intentional structure. It was so small that maturity was a more or less casual result of sporadic issuances. So that in 1923–24 20 per cent of the debt is floating. This is reduced to 13 per cent by 1929–30 and the marketable securities start to be spaced out a little more. But ten years later in 1939–40 there is a remarkable situation where 84 per cent of the debt is held in marketable securities with over 15 years to run and the rest is held in savings, and nothing whatsoever in floating debt or in securities under fifteen years.

TABLE 2.2
Maturity of the Irish National Debt 1923–24—1969–70
Percentages

Year	Floating Debt	Marketable Securities			Savings	External	Total
		−5	5−15	15+			
1923–24	20			73	7		100
1929–30	13		35	28	24		100
1939–40				84	16		100
1949–50	13		22	33	14	18	100
1959–60	29	4	27	18	11	10	100
1968–69	12	17	13	41	10	7	100
1969–70	11	21	13	38	10	7	100

The debt did not start to have much shape until the 1950's. By the end of that decade it is clear that a large quantity of short term debt had been created and this was causing the authorities some concern. Out of a total debt of £395 million, £117 million was in floating debt (see Table 2·1). The measure of the success of national debt policy is that by 1969–70, through funding, the quantity of floating debt is the lowest for any year given in Tables 2·1 and 2·2. However, this is only a temporary advantage, as 21 per cent of the debt is now (1969–70) held in marketable securities with less than five years to run. In absolute terms this means that £212 million of debt is due for redemption within the next five years. If this is combined with the floating debt, the government are going to have to recycle one third of the national debt within the next five years. This in a period when interest rates look as though they may be, on average, higher than they have been since 1945, is likely to form a major preoccupation of Irish debt policy. This preoccupation with day to day debt management is likely to be at the expense of an appreciation of the more subtle (and pervasive) effects of the debt on distribution, savings, and stabilisation policy.

The quantity of external debt as a percentage of the total has fallen from 1949–50 when it formed 18 per cent, to the present figure of 7 per cent; though this change in percentage terms represents an absolute change from £22 million in 1949–50 to obligations of £70 million in 1969–70. If government backed securities are included (line (b) Table 2·1) the distribution of the debt is not markedly altered, except that by 1969–70 the external debt was 9 per cent of the total rather than the 7 per cent shown in Table 2·2.

From these figures it is clear that on grounds of national debt management and, as will be shown later, in terms of stabilisation policy, the government will want to fund a substantial portion of the debt. There are many suggestions of ways to improve the marketability of government bonds each one of which suffers from its own peculiar disadvantages. A relatively new method which might fit the Irish case could be the issuance of 'strip' bonds on the American pattern. Strip bonds enable a purchaser to buy with a single issuance a whole range of different bond redemption dates. So that, rather like a Neopolitan cake, a purchase of a single bond would subsume two or three (or even more) different redemption dates and interest rates; thus with one issue the government spreads the load of its debt in a flexible manner, and the purchaser can fill a variety of gaps in his portfolio. Given the uneven spread of the Irish national debt such flotations might have attractions.

TABLE 2.3

	1	2	3	4	5	6	7	8
				2 as % of 1:	2 as % of 3:	1 as % of 3:	Consolidated public	2 as % of 7:
	Size of Debt £ million	Interest on Debt £ million	G.N.P. £ million	Interest as % of debt	Interest as % of G.N.P	Debt as % of G.N.P.	sector (current) £ million	Interest as % of public sector
1923–24	13·7	0·2	150★	1·5	0·1	9·1		
1929–30	24·3	2·1	161·4	8·6	1·3	15·1	34·3	6·1
1939–40	48·1	3·2	169·4	6·6	1·9	28·5	47·5	6·7
1949–50	121·5	6·7		5·5			108·7	6·2
1959–60	394·7	21·0	635·9	5·4	3·3	62·0	188·6	11·0
1969–70	1008·9	74·0	1438·0	7·4	5·2	70·0	411·6	18·0

★Estimate.

The cost of servicing the national debt is shown in Table 2·3. The Finance Accounts include in the service of the national debt 'Sinking Fund' provisions. The Sinking Fund provisions are not included in the figures for debt service in Table 2·3. These Sinking Funds are applied first, to redeem the debt, in which case the Sinking Fund is seen as a change in the capital stock and not as a current charge to service the debt; or secondly, are re-cycled through Ways and Means into the capital budget, in which case the Sinking Fund is again seen as a change in the capital stock and not as part of the debt service. So in neither case are the Sinking Fund provisions seen as a current service charge and for this reason they are not included in Table 2·3. As it is clear the government views the national debt as permanent rather than redeemable, sinking funds are an anachronism. As they are not used strictly to redeem the debt, nor should they be used as a backdoor method of providing Ways and Means Advances, they should be abolished. They only confuse an already confused situation.

The interest on the national debt includes not only that charged directly on the Central Fund and that included in the Issues for the Supply Services but also amounts applied to meeting interest on the national debt in the Capital Services Redemption Account. National debt interest as a percentage of the capital fell from 8·6 per cent in 1929–30 to a low of 5·4 per cent in 1959–60. (The figures for 1923–24 are included for interest, but reflect more the unusual times and are not really comparable to later figures.) Recently it has risen to 7·4 per cent (1969–70) and as one third of the debt is to be recycled in the next five years at interest rates which are likely to be higher than those in operation on the original issuances this interest charge is likely to rise still further.

It is interesting to note (column 5 of Table 2·3) that although the interest charge was 8·6 per cent on the debt capital 1929–30 this represented only 1·3 per cent of the gross national product; but the charge of 7·4 per cent in 1969–70 represents 5·2 per cent of G.N.P. Although the national product has been growing the national debt has been growing much faster and as a percentage of the G.N.P. (column 6 of Table 2·3) it has grown from 15 per cent in 1929–30 to 70 per cent today. Indeed, if the government backed bonds were included, the total debt backed by the government would rise to 82 per cent.

This increasing size of the national debt and the consequent increase in the service charge mean that a larger portion of the current public sector revenue is pre-empted to service the debt. Column 8 in Table 2·3 shows how the debt charge on the consolidated public sector current revenue has risen from 6·1 per cent in 1929–30 (and indeed was only 6·2 per cent in 1949–50) to 11 per cent in 1959–60 and 18 per cent in 1969–70. As will be discussed later, this has important fiscal and political implications.

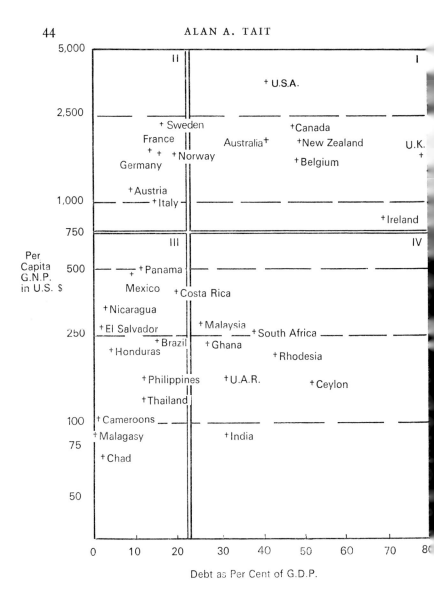

Debt as Per Cent of G.D.P.

Probably one of the best measurements of the size and importance of the national debt is to take it as a percentage of national
product and relate it to per capita income. Figure 2·1 shows the
national debts of 32 countries. The graph is divided into four
quadrants. In quadrant I Ireland is shown with a debt of 70 per

cent of gross domestic product and a per capita gross national product in U.S. dollars of $850. In the same quadrant the United Kingdom is shown as having the highest debt as a percentage of gross national product in the world at 79 per cent, but the per capita income of the U.K. is $1,620, double that of Ireland.

It is interesting to note the other countries in quadrant I of Figure 2·1; the U.S.A. (debt as a percentage of gross domestic product 41 per cent, and $3,520 per capita G.N.P.), Australia, Canada, New Zealand and Belgium represent (except Belgium) what General de Gaulle might have referred to as the white Anglo-Saxon countries. It could be argued that all these countries have taken their norms of debt behaviour from the standard set by the United Kingdom. That is, they have not regarded an increase in the national debt with the same aversion as other countries, and have been prepared to use the debt as a flexible part of total government policy.

This is in contrast to other European countries in quadrant II of Figure 2·1, Sweden, France, Norway, Germany, Austria and Italy, all of whom have comparable per capita incomes to those in quadrant I but who have much smaller national debts. This could reflect a different attitude to national debt creation. At the same time it might be more cynically remarked that many of these countries are those that have lost large scale wars, have suffered substantial disruptions of their currencies, have cancelled their debts, and started again. But a look at quadrant III and IV shows that countries with very low per capita incomes also fall into a pattern similar to that in quadrants I and II. The countries in quadrant IV are those countries with low per capita incomes who are, on the whole, territories where a British influence has been predominant, whereas in quadrant III other European or American influences have probably been more important. Once more those countries under the British attitude to debt policy are those who have the largest debt as a percentage of gross national product.

On this evidence, it could be held that Ireland by proximity, example, and a general tendency to imitate British financial policy, has been prepared to expand her national debt in a way which would not have happened had the British influence been less strong.

Figure 2·1 also shows it is impossible to talk of 'the biggest national debt in the world' in any meaningful sense. The United Kingdom has the highest national debt as a percentage of gross national product; on the other hand, Ireland has the second highest national debt as a percentage of G.N.P. and a per capita income roughly half that of the U.K. But Ceylon has a debt which is 52 per cent of the G.N.P. and a per capita income of only $150 per head (roughly one-sixth that of Ireland) and could be said to be in a much worse position than Ireland.

NATIONAL DEBT AND FISCAL POLICY

The accepted 'Musgrave' framework of fiscal policy separates the discussion into decisions affecting the allocation of goods and services, distribution, and stabilisation. Stabilisation subsumes questions on employment, price changes, growth and the balance of payments. It is interesting to fit Irish debt policy into this framework.

DEBT AND ALLOCATION

A government's main purpose is 'to get enough money to pursue their policies by means that do not cause them to lose power'.[1] Irish governments have an endless list of very worthy objectives which will cost money to achieve. Each expenditure can satisfy certain sections of the community and in so doing may 'buy' votes at the next election. On the other hand the taxes to finance expenditure are unpopular and tend to lose votes. The nice judgement is the mix of expenditure and taxes which will win the next election.

One way to weight the odds in favour of the government in power is to finance expenditure not by further taxation but by borrowing. If the government goes to the nation and borrows the cash necessary to buy its desired range of goods then the public makes a voluntary decision to buy the government debt. It is not forced to do so and it may simply exchange one form of asset (say, shares in a private company) for another (shares in the nation). As the purchase of debt is voluntary it must be presumed that the utility of the individual is increased when he buys bonds and that he considers himself better off.

[1] J. Wiseman, *Debt Management,* Institut International de Finances Publiques, Brussels, 1962, 335.

The current cost of debt issuance to set against the increase in well-being of the private sector is the taxation necessary to finance the interest payments and the eventual redemption of the debt. If information was complete and the public perfectly intelligent and sensitive, the disutility of the discounted future tax payments and debt redemption would be set against the increase in present and future satisfactions from holding the debt. It is widely held that the public is myopic about such transactions and concentrates on present benefits and underrates future costs. The result is the public feels better off where expenditure is financed by borrowing compared to the situation when the same expenditure is met by taxation. The temptation is created for governments to use borrowing as a politically more certain way of 'buying' votes than taxation.

It was this sort of reasoning which led to the 'sound finance' criterion of Victorian public accounting—that national debt was justified only to finance capital schemes. The capital asset then represented a tangible backing for the debt issue, but more importantly, this requirement put a limit on government 'vote buying' through debt creation. It meant the large portion of government expenditure categorised as current had to be financed by taxes whether this was unpopular or not.

Five circumstances have undermined this position. In the first place, wars have usually involved large government expenditure on current goods and services, not met by taxation. The borrowing to finance wars has created debt without the backing of capital assets (unless you count the survival of the nation as an asset backing the debt). In this way Britain built up the largest national debt per capita in the world (see Figure 2·1). Conversely, war-time can also provide the opportuinity for wiping the slate clean if assets are destroyed and the money supply has to be reorganised. This explains some of the low debt figures, even for highly developed countries, in quadrant II of Figure 2·1. Strictly, Ireland enjoyed neither of these experiences, but the so-called 'economic war' of 1932–38 did impose on her costs which were financed by borrowing e.g. $3\frac{3}{4}$ per cent Financial Agreement Loan 1953–58 for £10 million floated in 1938–39 represented over 20 per cent of the official debt at that time and yet did not have any asset backing and this was, effectively, the borrowing necessary to settle the economic war. To this extent it could be held the national debt policy broke away from the conventional sound finance attitude.

Secondly, the distinction between current and capital has become more blurred, and even the validity of that distinction is questioned. The more you read on the definition of capital the more you suspect the concept was originally formulated to justify limits on borrowing. You can hardly say that 'capital' was expenditure financed by borrowing, but if you define capital you can limit borrowing. So capital was associated with an asset which lasted longer than a year. If the asset lasted less than a year it could legitimately be financed from current revenue. But a year is an arbitrary accounting convention and there is nothing magic about it. Currently it is often argued that national accounts should be presented in a framework longer than a year.

Moreover, there are many 'assets' which are in an ambiguous situation, neither capital nor current. If the Irish government finances a scheme to eradicate bovine tuberculosis is this current or capital? Clearly the expenses are current but they are creating an increase in the value of a national asset—the livestock herd. (The Irish government in this case plumped for it as a capital expenditure.) If the State re-trains workers to meet changing regional and social patterns, is this capital or current? Indeed, is education generally capital or current? The expenses are mostly current (buildings are clearly capital) but the national stock of educated labour is improved. Conventional vote accounting highlights this difficulty. From the finance accounts for 1968-69 the State appeared to spend £106 million on capital goods and raised £114 million through national debt issues, leaving £8 million capital finance for current purchases; but from the national income accounts public authority capital expenditure was £131 million and capital receipts £95 million, leaving £36 million excess of capital expenditure to be financed from current revenue.

The third crack in the traditional mould is the actual means of debt flotation. When the public buys debt it is a voluntary decision. At that point in time they prefer an asset structure including government bonds, so this purchase increases their satisfaction. What happens if a government leans heavily on certain sectors of the public to 'encourage' them to take up government debt? Then the purchase is not entirely voluntary and the utility argument is questionable. Certainly Irish banks could refuse to buy the amount of the debt the Irish government suggests they take up, but there are unspoken suggestions that this might be unpatriotic and tactless. There is also the implication that the government finds it cheaper to finance debt issuance this way because interest rates would have to be higher if the whole debt was to be taken up by the non-bank public. Another way to say this is that the government has not been prepared to make the debt sufficiently attractive to ensure its total sale to the non-bank public.

The other argument is that as the banks accumulate the national savings and as company investment opportunities are limited it is 'efficient' for the government to mobilise the savings through bank subscriptions to a national loan. However, this has important implications for monetary policy which will be picked up later. For the time being it should be noted that the extent to which the government is successful in floating debt through government agencies and the banks, the discipline of the market place is slightly relaxed and the changes in satisfaction in the private sector do not occur. In fact the banks have been remarkably cooperative in buying substantial amounts of government debt. As Table 2·4 shows the Associated Banks have taken up from 32 per cent to 100 per cent of recent government issues. Even when the purchase of stock by the banks fell to 32 per cent this was at a time when the banks were already purchasing 45 per cent and 54 per cent of other government stocks.

TABLE 2.4
Government Stocks taken up by the Associated Banks 1966–70

1	2	3	4	5	6	7
		Total Nominal Issue £'000	Banks Bought			
ue Date	Stock		Conversion	New Money	Total (4+5)	6 as % of 3
/9/66	Exchequer Notes 1967–8	22,000	—	22,000	22,000	100
/3/68 /9/68 }	6½% Exchequer Stock 1971	24,879	8,537	7,800	16,337	66
/9/68	7% Exchequer Stock 1975	40,020	—	33,600	33,600	84
/9/69	8½% Conversion Stock 1970	18,194	3,158	6,125	9,283	54
/9/69	8½% Conversion Stock 1971	23,082	—	10,250	10,250	45
)/9/69	8½% Conversion Stock 1972	28,165	—	9,096	9,096	32
/3/70	8½% Conversion Stock 1973	41,885	8,810	25,285	34,095	81

In 1966 they took up the full £22 million new money issue and in 1970 81 per cent of the £41 million issue. These purchases must substantially bias the banks' portfolios and it is difficult to believe such amounts would have been bought freely. This means the government can issue more debt at lower rates than it could otherwise and thereby reduces its call on current revenue from taxation. But the non-bank public are not making the voluntary decision to buy all the debt so that the real cost of government policy is not brought home to the general public either in taxation or in bond purchase.[2]

This leads to the fourth and most important crack in the orthodox financial wall—the justification of many expenditures to stimulate national economic growth. As an economist wrote about the Irish debt in 1962 'an increasing debt may be a prerequisite for its proper contribution to the growth of the economy, evidence of a responsible, not irresponsible administration'.[3] This justification of debt finance has two branches, one in that there are limits to taxation beyond which incentives to work, save, invest, and take risk will damage the economic growth rate. The other is that in a developing country the state may have to take the lead in many forms of expenditure and this multiplicity of objectives may compel the state to finance current expenditures from borrowing. This policy issue then becomes one that the whole range of government expenditures is considered, then current tax revenue is offset against these expenditures, and the residual *has* to be met from borrowing.

Related to this, but separate, is the fifth circumstance affecting traditional views on 'sound finance'. Not only are governments held responsible for economic growth, there are also ever-increasing demands on them to expand their capital budgets. Traditionally, the capital budget could be considered as small and therefore demands on borrowing would be slight. Capital commitments nowadays are so large that in themselves they lend greater relative weight to the implications of borrowed finance.

Naturally, the danger these five cracks might expose the nation to is a flood of 'reckless' government expenditure. If war is

[2] The same argument applies to Land Bond issues which persons are obliged to accept in payment for compulsory purchase by the Land Commission.
[3] E. Nevin, *Public Debt and Economic Development*, Economic and Social Research Institute, Dublin, 1962.

accepted as an exceptional occurrence requiring exceptional expedients, the other four circumstances of lack of definition of capital items, encouraging banks to take bonds, the priority of growth over other objectives (e.g. 'sound finance') and the growth of the capital budget, still means that the old restraints on government 'buying votes' through debt finance have been relaxed. It means, if accepted, that the burden is thrown largely onto the probity of the government to ensure that, in the absence of any external criteria, it still conducts the business of debt finance with national responsibility rather than party bias. And this is a great deal to ask of politicians.

In this context it is worth noting that the two rapid, concentrated increases in the Irish public sector 1930-34 and 1947-51, were both due to increases in open ended expenditure on agricultural subsidies and housing, and were financed mainly by borrowing. Unlike other countries the rapid expansion was not sanctioned by public acceptance of increased taxation, but rather depended on the 'ability of the government to increase the national debt'.[4]

If the allocation of government expenditure between current and capital is no longer a strict function of borrowing, is there any other allocative function of national debt? It is still an important limitation of the finance of local authorities and semi-state bodies. Local authorities in Ireland are usually too small to involve themselves in the expenses of a bond issue. Expenditure is financed either through rates or subventions from the central government, and it is particularly true of three of the most important local authority areas of capital expenditure—housing, roads, and sanitary services. In 1968-69 the total outstanding debt to local authorities under the Housing Act was £60 million and under sanitary services £13 million, and the two together represented 8 per cent of the total national official debt. But the allocative decision remained with the central government.

Some semi-state bodies (nationalised industries) are in a stronger position than local authorities. The large companies (Aer Lingus, C.I.E., E.S.B.) can arrange their own flotations—although these are usually backed by the State.

---

[4] A. A. Tait and M. O'Donoghue, 'The Growth of Public Revenue and Expenditure in Ireland', in J. A. Bristow and A. A. Tait (eds.) *Economic Policy in Ireland*, Dublin 1968, 286.

The government through its national debt policy could strongly influence the allocation of funds between public and private sectors and within the public sector between central government, local authorities, and semi-state bodies. Interestingly enough some of the semi-state bodies are sufficiently large (and independent?) to take individual action if the State does not meet their capital requirements. Recently Aer Lingus has arranged external finance for aircraft purchase and there seems no reason why other companies could not arrange similar credit extensions. At one blow this reduced the allocation power of the central government and creates external liabilities which although not officially part of the national debt must be viewed by externals as guaranteed by the government, and to be as good as any other bonds backed by the Irish State.

In this way it is likely that the 'pecking order' for the allocation of capital funds within the public sector will be first the central government, then the large semi-State bodies, then the local authorities, and finally the small semi-State bodies who do not have the same claims as local authorities nor can they act independently of central government as can the large semi-State bodies.

Apart from allocation within the public sector the national debt has a huge impact on the allocation of command over total resources split between public and private sectors. If the government borrows it does so to get control over resources. To the extent that it succeeds it leaves less for the private sector. The total supply of savings over the last five years is shown in Table 2·5 and the government claim on it by new national debt (*all* new debt including savings, prize bonds etc.) is shown as a rapidly rising percentage from 50 per cent of public sector saving up to 84 per cent in 1966, and although falling in 1967 and 1968 has probably risen again in the most recent years.

Such claims by the State on the savings of the community are huge. It means that most of new savings of society are channelled by the State into investments which the government considers necessary, rather than allocated by the private sector. Obviously all governments claim part of new savings and what constitutes a 'proper' proportion cannot be defined; nevertheless, the proportion in Ireland is large by international standards and when it grows to over 80 per cent of the private sector savings available (1966) it might be reasonable to say that Ireland has ceased to be a private enterprise economy.

TABLE 2.5

	1 *Total Savings*	2 *Private Sector★ (personal & co.)*	3 *New National Debt*	4 *3 as % of 1*	5 *3 as % of 2*
1964	96	93	47	49	50
1965	107	103	62	54	58
1966	113	97	81	72	84
1967	141	124	56	40	45
1968	156	172	50	32	29

Source: *Statistical Abstract of Ireland 1968*, Prl. 189.; *National Income and Expenditure of Ireland 1968* Prl. 1064.
★Adjustment for stock appreciation attributed entirely from private sector.

This figure can be even more dramatic when it is considered that of new capital raised by issues of marketable securities in 1967 the State and State-sponsored bodies (nationalised industries) took 95 per cent. (This figure is much higher than the one quoted in Table 2·5 because retained profits provided much new investment and reduced the need for Irish companies to go to the market.) It is true that large amounts of government capital are recycled back to industry as capital grants and thereby reinvested in the private sector. This does not detract from the essential feature of this huge demand by the State on new savings, which is that it flows through the government and gives the disposition of the funds to government rather than to the private market.

To summarise, the allocation of goods and services between the public and private sector has been dominated by the demands of debt flotation; central government and large semi-state bodies are the agencies responsible for allocation of the resources raised through debt flotation, and traditional restraints on debt financing have been loosened by the definition of capital items, 'encouragement' on banks to take up new debt, and the priority of growth over other objectives.

DEBT AND DISTRIBUTION

The distributional impact of the size, structure, and administration of the national debt is rarely mentioned. We can consider the impact of the debt on the distribution of capital and of income.

When new debt is created it adds to the stock of government

capital but may not add to total capital. It may involve simply the transfer of savings which would have found a home in private sector assets and which instead are held in government bonds. Whether the creation of national debt adds to total savings or not may depend on the point of national economic development; if a country has an under-developed capital market and few asset forms in which wealth can be held then the availability of State debt, by giving an alternative home for savings, may increase total savings. In an advanced economy with a sophisticated capital market the national debt competes with numerous other forms of assets and its availability is more likely to involve a transfer between different asset forms rather than any net addition to savings.

The Irish position is unusual. There are a limited number of domestic investment outlets; the industrial base is small and in the hands often of closely held companies or companies held substantially by outside interests. Land, agricultural stocks, and property are the principle other asset forms available and it is the regrettable character of land and property that the supply is limited so that an increased flow of funds into those assets simply drives up their price without adding anything at all to the real national product. Therefore the temptation for an owner of new savings in Ireland is to invest them through the London stock market in assets outside of the State.

In these circumstances it is reasonable to argue that simply the availability of Irish government bonds, although not necessarily adding to total savings, has increased the flow of funds held domestically within the State.

New government debt must be issued at terms which make it attractive to investors. This means that its issue price and interest rate must be better than existing government bonds. In turn, this means that new debt makes old debt less attractive and as demand for the old bonds will be reduced the price of old existing bonds will fall. This fall in the price of existing bonds represents a 'tax' on the holder of these old bonds. Indeed, what is true of bonds is theoretically true of all existing assets; the issuance of new debt makes all existing assets less attractive, lowers their value, and hence is a tax on existing wealth. There is, therefore a redistribution from the holders of existing wealth to the owners of new wealth.

Against this 'wealth effect' there is the redistribution of income. New debt creates new obligations on the government to raise taxes to finance the interest payments, and eventually the redemption of the capital. This creates a transfer of the income flow from income receivers to bond holders. Bond holders are wealth owners and are few compared to the numerous taxpayers who finance the interest payments. This is likely to be, therefore, a redistribution from non-wealth holders to wealth holders.

In Ireland the savers of new wealth who invest in government stocks directly and indirectly may be relatively small savers. Figures could suggest substantial purchases of small amounts of national loans by the not-so-large savers. Of national bonds maturing 1966–77 only 17·5 per cent were purchased in applications of over £10,000; of the 7½ per cent exchequer stock 1973, 80 per cent was taken up by applications under £10,000; and the large (£19·4 million) issue of 3 per cent exchequer bonds 1965–70 was 78 per cent subscribed in applications under £10,000. In addition, large applications are often from the banks and this could represent the mobilisation of small deposits in the commercial banks. So although there is redistribution to holders of new wealth through issues and interest payments, in Ireland this could well be redistribution to new relatively small savers.

DEBT AND STABILISATION: EMPLOYMENT

The neo-Keynesian attitude to debt originated in the background of unemployment. It justified the use of debt formation to finance budget deficits to stimulate demand and expand economic activity. The transfer of income and wealth did not destroy any real resources. The use to which the funds were put could activate real resources, particularly labour.

It is often forgotten that the situation may be quite different in conditions of full employment; then debt financing cannot, by definition, activate any real resources because they are already fully employed. The resources the government gets can only be released from private consumption or investment.

In this instance a valid case can be made for borrowing abroad. By borrowing abroad no call is made on private capital or investment at the present. The resources borrowed abroad can be spent abroad on, say, capital equipment and used to promote a future stream of rewards to finance the future interest payments.

The interest payments will, of course, create a foreign claim on say, Irish output, but this could well be offset by the enhanced stream of domestic rewards from the investment.

Once more, the Irish case is unusual. Although we have continuous unemployment the problem is not the Keynesian one of stimulating demand. Ireland has managed to combine 6 per cent unemployment with over 6 per cent inflation in the last three years. Although we have substantial unemployment the mix of our resources results in price increases combined with unemployment. In Ireland today debt policy is better viewed in an anti-inflationary setting as a means of correcting structural deficiencies. There is the need to create new productive capacity without stimulating demand on domestic productive capacity. To that end, external borrowing to buy specific capital equipment could very well be a most valuable way to expand capacity with a minimum overspill into the domestic income stream (there would be some from transport costs, building etc.). This external borrowing will be discussed later in the context of the balance of payments.

DEBT AND STABILISATION: INFLATION

The anti-inflationary role of debt stabilisation policy in Ireland should be important. There are limitations on the freedom of fiscal policy (easy income and tax comparisons with Britain, trade union reference to cost of living claims, small income tax base etc.) and difficulties in conducting effective monetary policy (joint stock commercial banks operating both within and outside the State, dual currency circulating within the State etc.). It is possible that a thoughtful, continuous debt policy conscious always of the desired liquidity mix in the economy could be an important supplement to the somewhat ham-strung traditional fiscal and monetary policy. To date there has been little evidence of debt viewed in this light.

The crux of debt stabilisation policy is its influence on liquidity. Debt stretches from the completely illiquid such as irredeemable bonds (of which there are none in Ireland, unless you count as more or less irredeemable debt, capital raised by organisations such as the Agricultural Credit Corporation or Nitrigin Éireann Teoranta and guaranteed by the State—these are included in Table 2·1 line (b)) to the completely liquid which is money.

Between these extremes there is, potentially, every variety of liquidity. Because of the discontinuous nature of debt issuance not every gradation is possible in practice; but as the debt rolls forward, maturity is always changing and policy issues continuously change.

The most obvious policy issue is that it is never necessary to borrow. The government can always print money. When the government issues debt it means it has decided to do so in preference to issuing money (or indeed, to greater current taxation, or even to reducing expenditure). This must be a definite conscious decision, because creating debt imposes a cost on the exchequer which an increase in the money supply does not. Of course, many governments do not have complete control over the money supply. The assumptions that debt should consist of irredeemables only, with money satisfying everyone else, belongs 'in a Patinkinian world displaying assumed characteristics at which even the most ardent of science fiction devotees would boggle'.[5] However, the cost of the debt is deliberately undertaken by the government as a price to persuade persons not to spend on private consumption and investment but rather to put their funds into public goods. The issuance of money increases private demand for goods and services, and the creation of debt switches this to the public sector and in so doing alters the whole liquidity structure of the economy.

An efficient government liquidity policy is one that 'secures the desired degree of non-spending or illiquidity, at the least cost'.[6] This least cost depends on the correct assessment of requirements of the non-homogeneous mass of private lenders, the variety of non-homogeneous private debt competing with the public debt, and the mix of public debt to be issued.

Consider the question of whether selling debt would be deflationary. This is a more complicated question than it looks at first sight. It really should be re-phrased to whether debt issuance would be the most efficient deflationary policy. The private sector surrenders cash to buy bonds and thus becomes less liquid. But, in turn, this depends on four possibilities.

First, the cash may be realised from the sale of some other asset in the private sector. This would not be deflationary except

[5] E. Nevin, 'Debt Management: A General Report', *Debt Management,* Institut International de Finances Publiques, Brussels, 1962, 170.
[6] R.A. Musgrave, *The Theory of Public Finance,* New York, 1959, 582.

in so far as the sale forced down asset prices in the private sector and through a wealth effect reduced consumption. Quite apart from the assumption that private consumption alters as capital values change (rather than, say, savings), in Ireland this is less likely as the asset realised to enable the switch into bonds may well be external to the State. If an Irish resident sells U.K. securities to buy Irish bonds the wealth effect is felt in Britain and not in Ireland (except in so far as other Irish holders of U.K. stocks suffer the wealth effect). To that extent there is a 'liquidity leakage' in Ireland's case.

In the conventional analysis the deflationary impact of debt sales depends, in the first place, on the nature of the switch in the private sector. This then is further complicated in Ireland by the ease with which the switch can be made from external to internal assets and vice versa. There is no way to control this except by complete capital movement control, but the likelihood does diminish the first deflationary impact of Irish debt sales.

Secondly, as described above, the public will only be persuaded to buy the new debt if it is issued at rates and with conditions more attractive than existing debt, and this is likely to push down the price of the old debt. It is argued that this could first, reduce consumption due to a wealth effect; secondly, that as asset values fall persons are 'locked in' unless willing to realise losses and hence more illiquid. An extension of this argument claims that long-term debt will fall farther than short-term and hence create a greater locked-in effect; the policy recommendation from this is to fund the debt. As Musgrave says 'let us now set aside these considerations and assume rational behaviour by all concerned'[7]. When an investor suffers a loss, whether he sells security or not does not alter the existence of the loss. The only relevant question is whether it is more profitable to hold the security or to switch to some other. It is not the relationship to past prices which matters, but the expectation about future prices. And here there is a perverse factor; if holders of government debt believe that its price will fall further (the elasticity of yield expectations exceeds one) then they will be prompted to realise their loss and save themselves further losses. In this way the sale of new debt may force down prices and precipitate sales which help to bring

[7] *Ibid.*, 604.

about that which prompted investors to sell in the first place. So a self-fulfilling prophecy is created.

This fall in values would be deflationary, but it also creates problems for debt management. Although rising interest rates and falling bond prices are deflationary, a large and prolonged fall in government bond prices is bad for the economy. It is particularly so for an economy which wishes to increase economic growth and to that end wants to mobilise domestic savings through debt issues. A growing reluctance on the part of Irish investors to accept in an inflationary period increased amounts of government bonds each year may hint at an emerging problem in Ireland. If the government was prepared to quote completely market terms and stop 'leaning' on the banks to take government debt the resulting fall in existing debt prices might emphasise the problems outlined above.

Nevertheless, it is probably sensible for the Irish government to give some thought to British controversy over the support of an orderly market. In the next decade the Irish government is likely to continue as a seller of bonds and also be faced with considerable quantities of maturing debt for re-issuance. From Table 2·2 it can be seen that a third of the Irish debt (representing over £300 million) is due for redemption within the next five years. This is likely to be combined with a period of high interest rates and continuing inflation which will, in turn, require deflationary fiscal and monetary policies. 'At a time at which market expectations about future prices are unfavourable, we may well be unable to persuade the public to buy securities in sufficient volume to reduce the pressure on the banks which we would wish to see'.[8] But the government should always be able to persuade the public to take bonds provided it is prepared to face up to dropping the bond market with a bit of a bump. The recommendation then becomes the Radcliffe one of not necessarily supporting the long end of the market but of allowing it to move abruptly and then, at the new level, helping to run an orderly market. This is a far cry from the present Irish government bond policy (see below) but could become increasingly important if steps are taken to activate an Irish bond market.

[8]*Report of the Committee on the Working of the Monetary System,* Cmnd. 827, London, 1959, 26.

Thirdly, the deflationary impact of debt sales can depend on the type of private lender. There are numerous possible purchasers of government debt; an individual could well reduce consumption as he exchanges cash for bonds. However, it can be argued that as he achieves desired ratios of bonds, to equities, to cash, his future flow of income is more assured and his consumption may rise. This seems particularly likely where the country is relatively underdeveloped and the range of alternative investment opportunities for cash limited. Once more, the situation in Ireland is different from that in Britain; the capital market is stunted as few companies are quoted, dealings in the market move the price against the seller or the buyer, land and property offer the main alternative investments and the income from these is often more in the form of capital gains which require realisation for enjoyment than in a stream of income.

The result may well be that in Ireland purchase of bonds securing a guaranteed stream of income raises consumption rather than reduces it.

The main alternative to the individual purchaser of national debt is the banks. Once again, the sale of debt does not necessarily reduce inflation. Conventionally if banks use cash to buy bonds they become illiquid and reduce lending and hence private consumption. But this in turn depends on two further assumptions; first, whether the banks are operating at, or near, their cash and liquidity ratios. Clearly, if there was slack in these the bank could take up the bonds without changing its lending policies. The peculiarity of the Irish scene has been the 'rubber-like' quality of these constraints. If balances held with the Central Bank are regarded as cash, the cash ratio (see Table 2·6) has fluctuated between 8·7 per cent and 5·9 per cent in the last twenty years. If the Central Bank balances are considered call money the cash ratio has fluctuated between 8·4 per cent and 4·6 per cent. In either case considerable flexibility has been tolerated by both the banks and the Central Bank in the operation of cash ratios. The liquidity ratio shows a similar flexibility fluctuating between 16·8 per cent and 20·2 per cent over the same period. This indicates that for Ireland the cash and liquidity ratios have a much less restrictive character than in many countries. Of course, there is some minimum below which the banks would be unwilling to see their cash and liquid assets fall, but it is difficult to believe, given the variations in the past, that the banks would suffer a squeeze through cash or liquid asset changes if it clashed severely with bank lending policies.

TABLE 2.6

Irish Cash and Liquidity Ratios—Percent

| | Cash Ratios: Central Bank balance as | | | | | | Liquidity Ratio | | |
| | Cash | | | Call Money | | | | | |
	In State	Elsewhere	Total	In State	Elsewhere	Total	In State	Elsewhere	Total
1950	2·6	23·1	8·7	2·2	23·1	8·4	4·2	46·7	16·8
1956	2·7	16·0	6·5	2·5	16·0	6·3	7·4	48·4	19·1
1960	3·3	12·2	5·9	2·4	12·2	5·2	6·0	47·8	18·1
1965	6·2	12·9	8·1	2·4	12·9	5·4	13·0	33·1	18·8
1967	6·5	13·3	8·1	2·0	13·3	4·6	15·3	36·6	20·2

Moreover, there is no watertight way in which the Irish Central Bank can indulge in open market operations to influence the cash and liquidity ratios of Irish banks. For instance, if the authorities sold bonds to Irish banks these could be bought by realising assets held outside the State. The switch would therefore be between external and internal assets, but not necessarily between internal cash or liquid assets and internal bonds. Similarly if Treasury bills (Irish Exchequer bills) were sold to the banks the switch could be between external and internal funds and not necessarily change the internal liquidity structure. Obviously the more funds the authorities can pull back into Ireland the more effective will be their monetary policy. But associated with such a move there would have to be the establishment of accepted cash and liquidity ratios.

In the absence of these, 'Quantitative management of bank credit is implemented through the voluntary support by the associated banks of the aim specified in the letters of advice issued from time to time by the Central Bank'.

The further assumptions that sales of bonds to banks would be deflationary depends on the type of asset sold. This is the basis of the so-called 'new-orthodoxy' controversy in Britain. Part of the legacy of the Radcliffe Commission gave Treasury bills a central part to play in influencing the volume of bank's liquid assets, which in turn, set a ceiling to the expansion of bank deposits.

Thus the Central Bank by selling Treasury bills can reduce banks' cash and thereby their deposits and advances.

The evidence as to whether this is correct is conflicting and the clash of professional protagonists sounds impressive but there is no clear victor. Briefly, the opposition to the 'new-orthodoxy' holds that contraction of the Treasury bill supply is neither a necessary nor a sufficient condition for contraction in bank deposits, a contraction in cash is both necessary and sufficient for such contraction. It can be held that, say, sales of Treasury bills to the banks expand the liquidity base (the cash/Treasury bill ratio can be held by sales of long term assets) and further lending is facilitated. This might be summarised by saying that the deflationary leverage exerted by the Central Bank is greater when cash is mopped up than when Treasury bills are taken in. In Ireland there is no such controversy because open market operations are hampered, as mentioned above, by lack of accepted ratios, but also by the poorly developed market in government securities and the relatively small official holding of government debt.

Official holdings of government debt in September 1968 were about 16 per cent of the total outstanding bond issue; but over half of this consisted of non-marketable certificates of indebtedness under the Bretton Woods Agreement Act 1957. In effect the Central Bank could only deal in one stock (6 per cent funding loan 1969).

Finally, there remains the essential question whether debt sales are more or less deflationary than alternative policies (see Chapter I for further discussion of this point). There are at least three alternatives; government expenditure could be cut, taxation could be increased, or the money supply could be increased. We know that pound for pound reductions in government expenditure are more effective than cuts in taxation. So are debt sales more deflationary than taxation?

A direct tax reduces disposable income. The individual can react by reducing consumption, reducing saving, working harder or less hard, taking more or less risks and so on—all very thorny questions in themselves and mostly ones to which no unequivocal answer can be given. As far as the difference between identical amounts raised by direct taxes or debt, the main distinctions are largely distributional. As already mentioned, debt sales will reduce private asset values and it can be held that this effect is exactly the same as a direct tax on wealth. We cannot know. But whether the once and for all effect of an income tax on work effort will be different from the cumulative effects of direct taxes to finance the future stream of interest payments and eventual redemption, belongs to that realm of economic thought where economists try to estimate how many angels can dance on a pin-point.

The distributional characteristics of financing by debt or taxation will clearly be different. On the one hand debt creates assets which will be held by the private sector and which will increase the stream of income to a few people financed by many. On the other hand direct taxes pre-empt private consumption and private asset formation. Whether the net effect will be more or less deflationary depends on your assumptions. If price elasticity was very high amongst bond holders a small improvement in the rate of return of bonds would persuade them to take up a new issue; but the small rise in interest would result in only a minor fall in prices and it could be that the increased volume of the new assets would more than compensate for the wealth tax effect.

In this case the wealth effect could be inflationary compared to a direct tax on all wealth, or indeed compared to a tax on all forms of income. Indeed a tax on all forms of income would require tax repayments (or offset against other tax liabilities) on capital losses and this would, in turn, need higher tax rates on income to yield a replacement income which would be likely to have detrimental side-effects on work effort, risk and so on. A further distributional character suggests itself in that there could be off-setting improvements in the position of debtors, as creditors are impoverished by the fall in capital values.

There is one limit to using debt as an option to taxation. That is where the debt has grown as large as it can. That is where there is a limit to debt creation. Some commentators have suggested that this is a fast approaching problem in Ireland. Is there a limit to debt size? In theory there is nothing to stop a government expanding a debt to the point where interest charges are so large they pre-empt the entire national income. Indeed they could be larger than the national income and the limit would then be the speed with which taxes could be levied and bond interest paid. Just to state these ludicrous examples emphasises the sorts of constraints there must be on the debt size.

The main constraint must be the frictional costs of adminis-
tratoin. There must be a point where the army of bureaucrats to
administer debt is too large, where the disincentive effects of
taxation are a significant brake on the economy, and where the
distributional implications of bond interest and tax payment are
too regressive. None of these are likely to occur suddenly;
gradually the disadvantages of imposing further bond issues on
the country would weigh more heavily than the alternatives of
taxation, reduced public expenditure, or printing money. The
first sign of such strain would probably occur with the increased
restrictions on government freedom of action with the current
budget revenue imposed by the prior demand to meet bond
interest.

Obviously as the proportion of current revenue which has to
be used to pay debt interest rises governments are frustrated in
their plans for current expenditure. The trouble is that the interest
payments represent past government decisions which impinge on
present governments, and if the present government continues
to satisfy its ambitions in expenditure by further debt creation it
knows that the burden of interest payment will fall on a future
government. In this way there is a certain built-in recklessness to
expansion of the debt. As can be seen from Table 2·3 Irish debt
interest as a proportion of current government revenue (or more
broadly as a proportion of G.N.P.) has risen faster than the revenue.
This is not necessarily due entirely to an expansion of the debt.
In a period of rising interest rates the larger the quantity of short
term debt the quicker will interest charges increase. In the last
six years this has been the position of the Irish government and
it is likely to continue for some time, compounding the difficulty
of the rapidly increasing absolute size of the debt.

This would indicate that the government must start to contain
the expansion of the debt and expand tax revenues. Of course, a
fall in interest rates could relieve the situation.

Another less obvious but pertinent observation, is that the
government could have a built-in interest in a continuing inflation.
As the largest debtor in the country it is relieved of debt as
inflation continues; both direct and indirect taxation have a built-
in buoyancy of yield as money values rise and current revenue
rises. So on both the swings and the roundabouts the government
stands to gain from the inflation.

STABILISATION: GROWTH AND THE BALANCE OF
PAYMENTS

As already pointed out if debt is floated abroad it has implications for the balance of payments which many consider to be a
genuine form of burden. Although internal debt simply involves
a transfer of claims and liabilities which involve no real diminution
of resources, externally floated debt creates a claim on the real
product of Ireland which must reduce the amount available for
domestic consumption and investment.

The calculation of relative benefits between externally and
internally floated debt is more complicated. With an economy
like Ireland's where unemployed resources cannot be activated
simply through stimulating demand, then borrowing abroad
may be a way of obtaining capital in addition to that generated
internally. The trick lies in using the capital raised externally
without creating a large multiplier demand domestically. This
should be possible if loans floated externally are used to buy
equipment abroad; the income is then generated abroad and no
domestic multiplier effect occurs. The increase in the productive
capacity produces a future stream of rewards which help to pay
the foreign interest charge.

An interesting discounted cash flow exercise could be done on
these transactions because the profiles of the time flows are so
different. The capital and the income generated by the equipment
would have high present values whereas the eventual repayment
in 15 to 20 years time in foreign currency would be substantially
discounted. Moreover the choice of discount rate could be most
important. In this case, it would, presumably, be reasonable to
take the actual interest rate as a discount factor. Thus the capital
inflow would be discounted at the foreign rate of discount to
obtain the present value of the capital inflow. The stream of
interest payments abroad would be discounted at the foreign
rate but the stream of domestic income generated by the imported
capital equipment would be discounted as a benefit at the domestic discount rate. Thus the lower the foreign discount (interest)
rate the lower the present value of future costs, and the higher
the domestic rate the higher the present value of benefits. Clearly,
a cost benefit calculation on this basis could well favour foreign
flotations compared to domestic ones especially when external
interest rates were lower than the domestic rate. Of course the

whole analysis depends on assumptions of domestic full employment of resources and the use of the foreign loan to finance equipment purchases. Foreign loans used to expand domestic demand for domestically produced goods, e.g. housing, would probably be the most inflationary and costly form of debt financing, for not only are foreign claims created but current domestic resources are simply re-allocated, and a form of heavy money stimulates inflationary demand.

SUMMARY

Let us try to draw together some of the discussion. The Irish national debt stands at £1,009 million, or if you include government backed securities £1,192 million. This represents 70 per cent of G.N.P. (or 82 per cent). The structure of the debt is such that one-third of it must be re-cycled within the next five years. The volume of work involved in such a procedure is likely to limit the authorities' horizon and reduce even further their awareness, or interest in, the allocative, distributional, and stabilisation effects of the debt.

The Irish national debt size is in the tradition of countries which have been associated with British techniques and outlooks on debt creation. This is in marked contrast to countries on the Continent, or developing countries that have been associated with Continental countries.

The traditional 'sound finance' policy of limiting the debt issuance to cover the capital budget, has been undermined by the increasing confusion as to what constitutes a genuine capital expenditure, and by the urgent need to use the debt to expand economic growth. Moreover the real cost of a debt policy may not have been brought home to the public as some debt sales have not been in an entirely free market.

On both occasions when the public sector has expanded rapidly relative to the national income, the rapid increases have been in open ended expenditures (agricultural subsidies and housing) and at both times were debt financed. The conclusion is that the traditional restraints on responsible debt finance have been relaxed and no new criteria have replaced them.

There is the growing power of large semi-state bodies to create what amounts to national debt in their own name; this at one stroke often increases the external national debt and reduces the power of the central government.

The increasing demands of the government have limited the availability of new savings to be channelled to the private sector, but at the same time it can be forcefully argued that had the government not mobilised these savings in Ireland they would have been invested abroad.

External borrowing can be justified, in the circumstances of Ireland, specifically for buying capital equipment. Such borrowing should be much more closely linked to capital purchases, and care taken that this external borrowing does not allow a consequent relaxation of the constraints on borrowing internally.

To use the debt for anti-inflationary policy the government may have to face up to more abrupt changes in the bond market and a more rigorous interpretation of commercial bank cash and liquidity ratios. In any case the government should create a more active bond market and the public sector holding of government stocks should be more varied and used more flexibly.

There is no clear indication whether increased taxation should be used instead of new debt flotation. But the size of debt interest as a percentage of public sector revenue (18 per cent) implies a necessary constraint on further debt creation, which, if the government is intent on continuing to play a prominent role in expansion, implies an expansion of tax revenue. If tax revenue does not expand, and government debt must be contained, then the public sector as a whole will be under some constraint.

CHAPTER III

The Flexibility of Irish Taxes on Incomes

L. K. LENNAN[1]

INTRODUCTION AND SUMMARY
OF RESULTS

The principle objective of this paper is to estimate the sensitivity of the yield of personal and corporate income taxes with respect to changes in personal income and corporate profits respectively. This information is required for any analysis of the built-in flexibility of the tax system—i.e. the automatic dampening effect of taxes on fluctuations in national income, which must be allowed for when discretionary policy is being formulated.

The results suggest that for an increase of say £10 million in personal income £1 million of this increase will be syphoned off by personal income taxation assuming 1967–68 rates, approximately half of this tax being due in the year of the increase in income, the remainder becoming due in the following year. If the major part of this increase is attributable to wages and salaries as opposed to the category non-corporate profits, rents and dividends, the increase in tax revenue will be slightly higher. This is due to the inclusion in the latter category of agricultural profits on which taxation is negligible.

There was also found to be an additional factor which exerted a downward pull on the rise of tax receipts over the period studied, such that if tax rates and money income had remained unchanged there would have been a steady fall in tax receipts. This was found to be attributable to increases in the number of children allowed for under the income tax allowances. This

[1] The author is indebted to officers of the Research Department of the Central Bank of Ireland for their many helpful comments on a first draft of this article. Any views expressed are, however, the personal responsibility of the author and are not necessarily those held by the Bank.

increase is due to the fact that with rising incomes over the period studied, more family units which previously were below the exemption limit have come within the tax net and so have begun claiming hitherto unused children's allowances.

The time series method for estimating the personal income taxation functions was checked by means of cross-section analysis and similar results were obtained. Additional information which emerged from the cross section results was that for 1954-55, at that year's rates and allowances, the marginal rate of taxation for persons earning under £700 p.a. was 0·056 while that for persons earning over £700 p.a. was 0·335. Thus if the £10 million increase in personal income which was postulated above was solely attributable to persons earning under £700, £0·6 million would be the increase in tax revenue whereas if it had gone to those earning over £700, £3·4 million would go in taxation.

The statistical results for the corporate sector were not as conclusive as those found for personal income tax and the inferences drawn must be treated with some reserve. The results suggest that for an increase of £10 million in undistributed profits of companies £3 million of this will go to the Revenue Commissioners. (The period in which this additional tax revenue will become due is not clear from the equations but it is likely to involve a lag of between one and two years.)

The equations for personal income tax and corporate taxation can also be used for forecasting purposes. The error of the forecast for 1968-69 was about 5 per cent for personal income tax, the forecast for corporate taxation, using the first difference equations, being correct to the nearest million.

Our equations for personal and corporate taxation were finally introduced into a simple Keynesian model (including corporate savings and an import function) and resulted in a coefficient of 0·18 for the built-in flexibility of these taxes[2] (i.e. with constant tax rates the automatic response of tax yields with respect to income changes reduces those changes by 18 per cent).

[2] This figure must only be regarded as a rough indication of magnitude of built-in flexibility as the calculation was carried out by ordinary least squares analysis of a simple static nine equation model. The model and the estimates of the parameters are not embodied in the text but are available on request from the author.

In many economic equations the lack of definite knowledge about motivations and the lag structures of key variables makes it difficult to choose amongst the variables and lags which should be included in the equations. However, in the case of the tax equation the relevant rules are set out definitively, both tax rates and the tax base are given and all that has to be done is to construct a tax series by multiplying the tax base by the appropriate rate. Unfortunately it is not all that simple. The complexities of allowances and tax rates (the reduced rate having reappeared again last year) make the task more difficult as does the formulation of an appropriate relation between the economic aggregate with which we wish to associate the tax yield.

PERSONAL INCOME TAXATION

Were there no exemptions or allowances the National Accounts definition of personal income would correspond to taxable income and hence taxation receipts in the case where a single rate is applicable. However, from personal income must be deducted the components which are not taxable (e.g. most agricultural income and certain transfer payments), deductions that the tax law permits (e.g. insurance, first £70 of interest etc.), and personal exemptions which themselves depend on the structure of the population and proportion of earned and unearned income. A further problem is that in dealing with incomes and deductions in the aggregate, and not of a single household, the various amounts of allowances cannot strictly speaking be added and subtracted[3] but should be expressed functionally as:

$$PT = f(R, PI, NTI, D, PA, DIS)$$

where PT = personal tax yield, R = rate of tax, PI = personal income, NTI = non taxable income, D = allowable deductions for income tax purposes, PA = personal allowances, and DIS = a distribution of income variable. But even this is a simplification. To make matters worse some people pay tax on an instalment basis, some through deductions at source, different types of income fluctuate differently, and there are fluctuations in the proportion of exemptions and allowances actually claimed. But

[3] E. C. Brown and R. J. Kruizenga 'Income Sensitivity of a Simple Personal Income Tax', *Review of Economics and Statistics* Volume 41 (August 1959), 262–3.

the feature which must cause most trouble in the analysis of Irish income tax is that both persons and companies are included in the total income tax yield.

This essay cannot begin to analyse income tax in relation to all these variables as present published Irish data are inadequate for such a purpose. The most urgent statistical requirements as far as this sector of analysis is concerned are (1) an annual distribution of income survey, if possible related to consumption by income groups (2) tax paid by income groups and (3) an annual survey of earned income, average number of wives, children and dependent relatives and their distribution over the income scale. The necessity for these figures becomes more urgent as the years go by, as it must be remembered that where time-series analysis is involved at least 14 or 15 observations would be required to give any sort of reasonable result. Thus even if these figures were published this year it would be some time before any major analysis could be attempted.

What is calculated in this paper is a composite action parameter —the aggregate marginal effective rate of tax. Before going into the detailed calculation of this rate it is as well to set out the theoretical basis initially. Assuming that PT = aggregate tax yield, PI = total personal income, and T = time, the statistical model will take one of the following forms.[4]

(1) $PT_t = f(PI_t)$

(2) $PT_t = f(PI_t, PI_{t-1})$

(3) $PT_t = f(PI_t, T)$

(4) $PT_t = f(PI_t, PI_{t-1}, T)$

Where (1) relates personal tax yield to personal income in that year, (2) takes account of the fact that along with tax payments for the year in question PT will include a certain proportion of tax receipts which are from people who are being assessed on their income for the previous year; (3) is similar to (1) but in addition to this it brings in a composite variable which is included

[4] E. T. Balopoulos, *Fiscal Policy Models of the British Economy*, Amsterdam 1967, 15.

Year ending	Changes in Taxes on Personal Income & Companies	Changes in Allowances Effect on Revenue		
		Current Year	Following Year	2nd Year
1954	———			
1955	Personal allowances increased, incomes below £240 exempted from income tax, income tax on residential property reduced.	−0·4		
1956	Tax free allowance for children increased.	−0·1		
1957	First £25 saving interest exempted from tax.	−0·05		
	Dependent relative allowance.	−0·106	−0·287	
	Introduction of 20% initial allowance for wear & tear.		−0·2	
1958	Tax relief for person making annual payments to University for research.	−0·015		
	Accelerated depreciation allowance.		−0·4	
1959	———			
1960	Introduction of wear & tear allowance on buildings and rebates for training local staff.	−0·015		
	Relief for purchased life annuities.	−0·010		
	Surtax limit and allowances increased.	−0·160		
	Standard rate of tax reduced.	−1·015		
1961	Exemption of profits of Harbour Authorities.	−0·07		
	Child allowance increased.	−0·11		
	Savings tax allowance extended to both husband & wife	−0·02		
	Extension of dividend relief for certain securities.	−0·015		
	PAYE 1 months payment lost because of payment in arrears.	−0·677[e]		
1962	Earned income relief increased.	−0·080		
	Surtax increases in rates more than balanced by increased exemption limit	−0·1		
	Other income tax reliefs	−0·02		
	Reduction of standard rate.	−1·2		
1963	———			
1964	Increase in Corporation Profits tax.	3·0		
	Rents to be based on actual income.	0·4		
	Anti-evasion measures.	0·6		
1965	New allowance for age introduced.	−0·18		
	CPT reduced.	−0·04		
1966	———			
1967	Standard rate increase	4·55		
	Allowances increased.	−0·35		
1968	Relief for illness.	−0·25		
	Child & dependent relative allowances increased.	−0·75		
	Surtax earned income relief increased, rates increased: balance.	−0·15		
	Initial allowance increased to 50%		−0·5	−1·0
1969	Married allowance increased.	−0·15	−0·19	
	Surtax earned income allowances increased.	−0·12		
	CPT deduction for directors' salaries.	−0·05		
	Initial allowances increased to 60%.		−0·55	−1·14

TABLE 3·1

Changes in Standard Rate Revenue Effect	CPT Changes	Surtax	Total Income Tax Changes		Income Tax Personal at Previous years Rates	Surtax at Previous years Rate
			Personal	Corporate		
			−0·4		14·7	2·1
			−0·1		15·0	2·1
					14·6	2·0
			−0·2			
			−0·2	−0·2	15·3	2·2
				−0·4	16·4	2·2
					17·2	2·1
015		−0·2	−0·8	−0·3		
			−0·8	−0·1	20·5	1·7
					22·1	2·0
		−0·1				
2			−1·0	−0·3		
					24·4	2·2
	3·0					
			0·7		26·2	2·2
			−0·2		33·2	2·6
	−0·04					
					39·8	2·6
55			3·1	1·1	42·2	3·2
					56·3	3·3
			−1·0			
		−0·15				
			−0·15	−0·5	64·3	3·5
	−0·05	−0·1				

= estimated as 1/5 of the five months revenue for 1960–61 which was £3,387,472.

to take account of various factors which happen to be correlated with time (e.g. income distribution, population movements etc.); (4) is similar to (2) with the additional refinement of a time trend.

The form which these equations will take will depend on whether one prefers to postulate a constant marginal effective tax rate or a constant effective tax yield elasticity with respect to aggregate personal income. For instance, the statistical model $PT_t = f(PI_t)$ can take the form $PT_t = t_o + t_1 PI_t$ if one chooses to postulate a constant marginal tax rate, or the form $PT = t_o PI_t$ if one prefers the hypothesis of a constant tax yield elasticity.[5]

As Balopoulos[6] shows t_1, is inefficiently estimated in any situation where the tax structure or the pattern of income distribution changes, unless the existing interrelationship between aggregate personal income and the other explanatory variables incorporated in the residuals remain unchanged over time. In order to avoid the first defect the tax yield series, following Prest,[7] is adjusted in this paper to account for changes in tax rates and allowances and although such adjustments cannot but be slightly crude they do go some way towards meeting this objection. The introduction of the time trend allows for the effects of the factors which may be correlated with time.

TIME SERIES ESTIMATES

The calculation of the marginal rate of personal income tax and surtax by time series methods is based on data for the fourteen fiscal years 1954-55—1967-68, and the method used where possible is that adopted by Prest[8] to enable comparisons to be made between the Irish and British results.

Due to the aggregation of corporate and personal income tax in the Revenue Commissioners' statistics, use was made of the National Income and Expenditure Accounts figures for taxes on personal income from which total social insurance contributions were deducted. These figures were then broken down into income tax and surtax by deducting the figures for net receipts of surtax published in the Annual Reports of the Revenue Commissioners.

[5] *Ibid.*
[6] *Ibid.*
[7] A. R. Prest 'The Sensitivity of the Yield of Personal Income Tax in the United Kingdom' *Economic Journal* (September 1962), 578.
[8] *Ibid,* 576–96.

The resulting data had then to be adjusted to give a series of yields at constant tax rates. The predictions given each year in the Budget Statement for changes in yield due to changes in tax rates, changes in allowances etc., were deducted from the total change in taxation to give a picture of the automatic increases. Thus it was assumed that the Revenue Commissioners had perfect foresight and that the increase in revenue over the estimate due to tax rate changes was due to income variations. No adjustment was made therefore for years in which income was inaccurately predicted. Where changes in tax rates were taken during the year, the excess of the increase in a full year over that in the current year was deducted from the following year's figures. Where changes in the income tax rates were involved the distribution of the effects on personal and corporate incomes was assumed to bear the same relation as the ratio between taxation on persons and taxation on companies. An analysis of the changes in taxation over the period and amounts involved is contained in Table 3·1.

The result of these calculations was the compilation of a series of yields which give comparable sets of data for any two successive years—i.e. the change from the preceding year if that year's rates had remained constant. This data was then converted to constant 1954–55 and 1967–68 rates[9]. The same adjustments were made to the surtax data and the two were added together to produce the series in Table 3·2.

Various equations were fitted by ordinary least squares using tax yield as the dependent variable and current and lagged personal income[10] as the independent variables. The following results were obtained (throughout, an asterisk indicates a coefficient the value of which is significantly different from zero at the 5 per cent level).

[9] See Prest *op. cit.* 579.

[10] The data for personal income do not correspond with those published in the national accounts as the figures were adjusted to correspond to the financial year—for methods of adjustment see Appendix. A more appropriate base for income tax would have been personal income less current transfer payments and agricultural income as fluctuations in these magnitudes are unlikely to influence tax receipts. Tests were run for this base resulting in a slightly improved statistical fit and higher marginal coefficients. However such disaggregation was not envisaged in the aggregate model of built-in stability of which this study is part and is not therefore investigated in this paper.

TABLE 3·2

Standardised Personal Income tax and

Surtax (£ million)

	1954–55 Rates	1967–68 Rates
1954–55	16·4	16·0
1955–56	17·1	16·8
1956–57	16·7	16·4
1957–58	17·8	17·6
1958–59	19·2	19·0
1959–60	19·9	19·7
1960–61	24·1	23·7
1961–62	27·1	26·3
1962–63	31·3	29·7
1963–64	33·5	31·4
1964–65	41·2	37·8
1965–66	49·1	44·2
1966–67	52·6	47·1
1967–68	64·5	58·4

1954–55 RATES

(1) $PT_t = -35.790 + 0.11951PI_t - 1.30301T$
$$(0.0124)^\star \qquad (0.5106)^\star$$
$R^2 = 0.9868 \quad SE = 1.941 \quad DW = 1.795$

(2) $PT_t = -35.800 = 0.11958PI_{t-1} - 0.71372T$
$$(0.0111)^\star \qquad (0.4068)$$
$R^2 = 0.9891 \quad SE = 1.764 \quad DW = 2.216$

(3) $PT_t = -37.2621 + 0.05371PI_t + 0.07017PI_{t-1} - 1.13022T$
$$(0.02691) \qquad (0.02665)^\star \qquad (0.41694)^\star$$
$R^2 = 0.9922 \quad SE = 1.565 \quad DW = 1.574$

(1a) $\Delta PT_t = -0.00150 + 0.094914\Delta PI_t$
$$(0.02604)^\star$$
$R^2 = 0.5250 \quad SE = 2.473 \quad DW = 2.427$

(2a) $\Delta PT_t = -0.00068 + 0.102754\Delta PI_{t-1}$
$$(0.027166)^\star$$
$R^2 = 0.5438 \quad SE = 2.446 \quad DW = 1.954$

(3a) $\Delta PT_t = -1.636102 + 0.06864\Delta PI_t + 0.07585\Delta PI_{t-1}$
$$(0.01987)^\star \qquad (0.02114)^\star$$
$R^2 = 0.7812 \quad SE = 1.759 \quad DW = 1.557$

1967–68 RATES

(4) $PT_t = -26\cdot661 + 0\cdot09718PI_t - 0\cdot87067T$
   $\quad\quad (0\cdot01177)^\star \quad (0\cdot48596)$
   $R^2 = 0\cdot9839 \quad SE = 1\cdot847 \quad DW = 1\cdot696$

(5) $PT_t = -26\cdot526 + 0\cdot09682PI_{t\,-1} - 0\cdot37663T$
   $\quad\quad (0\cdot01177)^\star \quad\quad (0\cdot41039)$
   $R^2 = 0\cdot9851 \quad SE = 1\cdot779 \quad DW = 2\cdot065$

(6) $PT_t = -27\cdot7943 + 0\cdot04657PI_t + 0\cdot05397PI_{t\,-1} - 0\cdot73774T$
   $\quad\quad (0\cdot0285) \quad\quad (0\cdot0282) \quad\quad (0\cdot442)$
   $R^2 = 0\cdot9882 \quad SE = 1\cdot658 \quad DW = 1\cdot584$

(4a) $\varDelta PT_t = 0\cdot13157 + 0\cdot080830\varDelta PI_t$
   $\quad\quad\quad (0\cdot02437)^\star$
   $R^2 = 0\cdot4779 \quad SE = 2\cdot315 \quad DW = 2\cdot237$

(5a) $\varDelta PT_t = 0\cdot22795 + 0\cdot084732\varDelta PI_{t\,-1}$
   $\quad\quad\quad (0\cdot026267)^\star$
   $R^2 = 0\cdot4640 \quad SE = 2\cdot352 \quad DW = 1\cdot816$

(6a) $\varDelta PT_t = -1\cdot19125 + 0\cdot05957\varDelta PI_t + 0\cdot06138\varDelta PI_{t\,-1}$
   $\quad\quad\quad (0\cdot02116)^\star \quad\quad (0\cdot0225)^\star$
   $R^2 = 0\cdot6885 \quad SE = 1\cdot873 \quad DW = 1\cdot475$

Where PT_t = personal income taxation in financial year
   $\quad\ PI_t$ = personal income in financial year
   $\quad\ PI_{t\,-1}$ = personal income (financial year) lagged one year
   $\quad\ T$ = time

All the coefficients in the equations have the expected signs and no negative income coefficients appear in the equations containing both current and lagged personal income as was the case with the British results. The R^2 values are high in all the equations and the DW values are significant. There is little to choose between the current and lagged income equations if they are judged by R^2s. However, the standard errors in the case of the lagged equations are lower and this together with the fact that one would expect to find a closer connection between tax and previous year's income (PAYE having only been introduced in the early sixties) must lead one to place more weight on the lagged equations.[11]

The equations suggest that the long-term marginal rate of personal income taxation in Ireland is of the order of 0·120 at 1954–55 rates and 0·097 at 1967–68 rates and the short-term rates are approximately 0·054 at 1954–55 rates and 0·047 at 1967–68 rates. If the equations specified above are tested for their forecasting power they give better results than forecasts based on the various other 'naive' hypotheses set out for testing by Prest[12] but are nevertheless on average 5 per cent lower than the recorded value for 1968–69.

To improve the forecasting power of the equations various other regressions were tested. The first study omitted the time trend and gave coefficients at 1967–68 rates of 0·077 for the current income equation and 0·087 for lagged income. However, the forecasts for 1968–69 disimproved and this factor together with the spread between the two coefficients led to the rejection of these equations. An additional regression carried out was of tax yield on time and a compound measure of income (CI) which weighted lagged and current income by their relative importance in total tax receipts. It was assumed that prior to 1960–61 all tax was paid on the basis of previous year's income. 1960–61 was weighted as between lagged income and current income (PAYE was introduced on 6 October 1960) and the remaining years were weighted by reference to the published statistics for PAYE as a percentage of total tax receipts. The coefficients were 0·092 at 1967–68 rates and 0·115 at 1954–55 rates but the forecasts, although better than those for the equations without the time trend, were not satisfactory. The first difference equations however were statistically superior to (1a) (2a) (4a) and (5a) and also reinforce the estimates of the long-term marginal rate of personal income taxation given earlier. They were:

1954–55 RATES

$$(7) \quad \Delta PT_t = -0·67653 + 0·116421 \Delta CI$$
$$(0·02648)^\star$$
$$R^2 = 0·6373 \quad SE = 2·139 \quad DW = 2·058$$

[11] When surtax is deducted from the taxation figures the current income equation gives better results.

[12] Prest, *op. cit.* 582.

1967–68 RATES

(8) $\Delta PT_t = -0.40205 + 0.097456 \Delta CI$
$\qquad\qquad\qquad (0.02632)^\star$
$\quad R^2 = 0.5544 \quad SE = 2.125 \quad DW = 1.884$

An additional study was carried out to ascertain the marginal rates of taxation on wages and salaries on the one hand and non-corporate profits, dividends and interest on the other. 1954–55 was omitted from the data because a breakdown of assessments by schedules was only available from that year and as the figures were for assessments they relate on the whole to tax paid in the following year. The division between tax paid currently and in arrears was made on the basis of the details of Schedule E Tax Payable since 1960-61 contained in the Annual Reports of the Revenue Commissioners. For earlier years the table 'Assessments under Schedule E' was used and these figures then converted to receipts by assuming that all the assessments carried out were paid in the following year[13] (this was done because receipts figures were required to correspond to the earlier series of total tax paid at 1967–68 rates). These data were then processed to give a series at constant 1967–68 rates as before: the adjustments for the effects of budgets, however, were a good deal more rough and ready as no breakdown was available for the estimated effects of changes in rates on Schedule E receipts.[14] The corrected series of tax yields at 1967–68 rates on non-wage incomes was obtained by subtracting the wages and salaries corrected series from the 1967–68 series in Table 3·2. The division of personal income into wages and salaries and profits, rents and dividends was made on the basis of Table 2 of the National Income and Expenditure Accounts of 1968. Salaries and wages were equated with remuneration of employees and the remainder of personal income was regarded as profits, interest and dividends. The following results were obtained:—

Wages and Salaries

(9) $WST_t = -14.095610 + 0.09880 WSI_t - 0.78223 T$
$\qquad\qquad\qquad\quad (0.01445)^\star \qquad (0.45678)$
$\quad R^2 = 0.9768 \quad SE = 1.664 \quad DW = 0.812$

[13] Although it might not be true in all cases that the assessments in one year are paid in the next this is the most appropriate general assumption.

[14] The changes in rates were allocated as between wages and salaries and profits and rents by reference to the ratio of tax paid on wages and salaries to total personal taxation.

(9a) $\Delta WST_t = -0.6572 + 0.10273\Delta WSI_t$
$$(0.0224)\star$$
$$R^2 = 0.6563 \quad SE = 1.457 \quad DW = 1.796$$

Profits, Rents etc.

(10) $PRT_t = -16.51843 + 0.11562PRI_{t-1} - 0.34144T$
$$(0.02164)\star \qquad (0.34508)$$
$$R^2 = 0.9703 \quad SE = 1.205 \quad DW = 2.799$$

(10a) $\Delta PRT_t = 0.1640 + 0.0811\Delta PRI_{t-1}$
$$(0.0500)$$
$$R^2 = 0.1897 \quad SE = 1.969 \quad DW = 3.005$$

The DW statistics for these equations are poor. Other variables were included in the equations but the serial correlation could not be eliminated. However, the equations do suggest that if the first difference equation for wages is taken as an indicator of the marginal rate of taxation on wages and salaries (there is no evidence of autocorrelation in this equation as the DW statistic is 1.796), the marginal rate is higher on wages and salaries in Ireland than on profits and rents. This can be explained by the fact that agricultural income accounts for nearly half of the income from profits, rents dividends etc. When agricultural income is omitted from the equation, a marginal rate of 0.1303 is obtained. (The R^2 and DW statistic are slightly superior in this equation.)

The importance of the surtax paying classes was ascertained from a regression of income tax at constant 1967–68 rates as before (excluding surtax) on personal income. The following results were obtained:

(11) $(PT-S)_t = -25.1833 + 0.08930PI_t - 0.79202T$
$$(0.00956)\star \qquad (0.40895)$$
$$R^2 = 0.9874 \quad SE = 1.575 \quad DW = 1.754$$

(12) $(PT-S)_t = -25.533 + 0.09001PI_{t-1} - 0.33765T$
$$(0.01058)\star \qquad (0.39722)$$
$$R^2 = 0.9851 \quad SE = 1.710 \quad DW = 1.932$$

(13) $(PT-S)_t = 25.778 + 0.05983PI_t + 0.03111PI_{t-1} - 0.69286T$
$$(0.03735) \qquad (0.03808) \qquad (0.43307)$$
$$R^2 = 0.9881 \quad SE = 1.601 \quad DW = 2.276$$

Comparing the coefficients with those in the earlier equations we find that the differences are 0·0079, 0·0068 and 0·0096 respectively. Although the standard errors are quite high, it is significant that all the results are about the same indicating that surtax contributes about one percentage point to the marginal rate of taxation which is similar in magnitude to the British results and shows that surtax contributes a larger proportion of the built-in flexibility in the Irish personal tax system.

What accounts for the difference between the marginal rate of tax at 1967–68 rates and at 1954–55 rates? It is known from official data that the average rate of tax in 1954–55 (at 1954–55 rates) was 3·8 per cent and that the average rate of tax in 1967–68 (at 1967–68 rates) was 6·5 per cent. The calculations of tax yield in terms of constant rates suggest an average rate of tax in 1967–68 (at 1954–55 rates) of 7·1 per cent and an average rate of tax in 1954–55 (at 1967–68 rates) of 3·7 per cent. Therefore the average rate of tax at 1967–68 rates was lower than that based on 1954–55 rates thus suggesting a small difference between the marginal rates on the same two bases. (It should be noted however that the difference is nowhere near as great as that shown in the British statistics as this was due to a high rate of taxation in 1948–49 which had not been reduced from the war years' high level.)

A comparison of the structure of the rates and allowances for Ireland in 1954–55 and 1967–68 does not throw very much light on why the differences should occur, as the increases in allowances would appear to have been cancelled out by the abolition of the reduced rates (see Table 3·3). However, when Table 68 of the Forty-fifth Annual Report of the Revenue Commissioners (Income Tax, Gross Income, Exemptions, etc.) is examined it can be seen that the ratio of the net produce of income tax to taxable income increased from 30·7 per cent in 1954–55 to 33·8 per cent in 1967–68 indicating that the changes in rates accounted for an increase of £6 million in taxation receipts. However, just taking the married, single and widowed allowances alone, the increases of nearly £100 in these over the period, applicable to approximately half a million tax payers, would have had the effect of lowering taxable income by £50 million and hence taxation by £16 million, thus more than compensating for the increase in rates.

TABLE 3·3

Rates and Main Allowances

		1954–55	1967–68
Standard rate		7/6	7/–
Higher reduced rate		6/–	—
Lower reduced rate		3/–	—
Earned Income allowance		1/4 of 1st £800	1/4 of earned income
		1/5 of balance (Max. £400)	(Max. £500)
Personal allowances:	Married	£300	£394
	Widowed	£175	£259
	Single	£150	£234
Children under 11 Children over 11 and under 16 and certain other.		£85	£135 £150

Source: *Forty-Sixth Annual Report of the Revenue Commissioners,* Prl. 734
       Stationery Office (1969) and earlier issues.

With regard to the negative time trend in the equations, an investigation was made into the extent to which this was caused by increases in the number of children's allowances.[15] Initially the number of children claimed for under the income tax children's allowance in 1954–55 was compared to that in 1967–68. The 1954–55 figure was simple enough to calculate as the allowance was constant for all children at £85. According to the Revenue Commissioners Annual Report the deductions in 1954–55 in respect of children were valued at £7,693,000 and this, taken with an £85 allowance, would suggest that 90,500 children were claimed for. The 1967–68 calculation is more complicated as two rates were in force, £135 for children under 11 and £150 for children over 11 and not over 16 and children over 16 in certain circumstances (mostly those receiving further education). On the assumption that all children who receive the allowance

[15] This is one of the possible explanations put forward by Prest (*op. cit,* 594) for the emergence of a negative time trend.

were under 11 the amount of children allowed for would have been 219,000 and if all the children allowed for were over 11 the number of children receiving the allowance would have been 197,000. There is also, of course, the problem that if the child is earning over £80 only a part of the allowance is paid, but this problem is ignored. In order to narrow down the limits, calculations were made using the 1966 Census of Population. Volume V, Tables 5a and 5b provided details of children between 16 and 20 receiving full-time education—no figures were available for those aged 20 but a figure of just over 4,000 was projected. This gave a figure of 60,000 for those receiving full-time education. The number in the age group 11–15 was then calculated from Table 1a Volume II (as the age groups did not correspond with the years required it was assumed that the number of children in each age group was distributed evenly over the years included) giving a figure of 350,000. Similar computations were used for the 0–10 year olds and the number of children worked out at just over 650,000. It was further assumed that the number of children in the taxation deduction class bore a similar distribution to those in the population as a whole and the allowances were weighted to reflect the magnitudes in each group. The average allowance worked out at £144·8 thereby giving a result of 204,000 children allowed for in 1967–68.

The increase in the number of children in respect of whom allowances were given is striking when compared with the small increase in the total number of children during the period 1951–66 (from 855,000 to 900,000 or 5 per cent). The allowance figures would indicate, if the increase was spread over the period, that 8,750 extra children were allowed for each year. At, say £100 per child the extra allowance would work out at £875,000 p.a. thus accounting for the major share of the negative time trend which, where it is significant in the equations, indicates a yearly negative drag in the region of £1·1—£1·3 million. The remainder of the negative time trend is probably accounted for by increases in the numbers claiming other allowances. For example, an analysis of the married and other allowances shows that, whereas in 1954–55, 80,000 persons claimed married allowances, in 1967–68 this figure had risen to 176,000 persons. The same trend was evident on the allowances for single and widowed people, the number of claimants more than doubling.

CROSS-SECTION ESTIMATES

In order to check on the validity of the earlier estimates of the marginal rate of tax, a cross-section approach was used. Unfortunately there are no published figures in Ireland for the distribution of personal income by ranges before and after tax. However, in a paper to the Statistical and Social Inquiry Society of Ireland, Reason[16] produced income and taxation figures by ranges for non-agricultural occupations for 1954–55. The basic material needed is set out in the Appendix Tables I and II of Reason's paper. Table I gives income classified by income ranges and the number of people in each range from which average income per person in the ranges can be ascertained. Table II gives average tax payable per person in that range. Various multiple regression analyses were carried out and the following was the best fit.

$$\log PT = -5{\cdot}8358 + 2{\cdot}6059\log PI \qquad (R^2 = 0{\cdot}9789)$$
$$(0{\cdot}0990)\star$$

This means that the elasticity of tax yield with respect to income was just over 2·6 in 1954–55. Translating this into marginal tax rates payable by each group (elasticity multiplied by mean tax paid over mean income per family)[17] gives:

£	£	
240–	300	0·0252
301–	350	0·0347
351–	400	0·0404
401–	450	0·0465
451–	500	0·0747
501–	550	0·1118
551–	600	0·1362
601–	700	0·1373
701–	800	0·1866
801–	900	0·2014
901–1,000		0·2013
1,001–1,100		0·3152
1,101–1,200		0·3448
1,201–1,300		0·3831

[16] L. Reason, 'Estimates of the Distribution of Non-Agricultural Incomes and Incidence of Certain Taxes' *Journal of the Statistical and Social Inquiry Society of Ireland,* Volume 20, 1960–61.
[17] A. R. Prest *op. cit.* 587.

£ £
1,301–1,400   0·3975
1,401–1,500   0·4295
1,501 and over 0·7080

When these were weighted by the total income received by each group (inclusive of the group below £240) the result was 0·117. This estimate looks reasonably close to the original estimate of 0·120 obtained by time series analysis. But account must be taken of cross-section bias which tends to inflate the estimate obtained by cross section methods.[18] This can be done crudely by assuming that the bias is similar to that inherent in the British estimates. If the same bias is present here as that in Balopoulos's estimate it would be necessary to increase the above result by 18 per cent giving an estimate of 0·138 which is a good deal too high for comfort. However, the agricultural sector is excluded from our cross section calculations and this is likely to exert a downward drag on the aggregate marginal tax rate if included.

An attempt was made to integrate the agricultural sector into the distribution of income and taxation framework. As no figures were readily available for agricultural income per person or taxation per person, rough estimates had to be used and this must be taken into account when studying the results.

The National Farm Survey on the distribution of family farm income per farm in the State (Table B Chapter 5) was used to break down the item 'income from self-employment and other trading income' for 1954–55 in Table B1 of the National Income and Expenditure Accounts 1968. In order to translate the ranges of income given in the Farm Survey into those compatible with the distributions in Reason's paper it was assumed that incomes were spread evenly within each income range. The following were the total income and number of persons in each range:

TABLE 3·4

Under	(1) Income (£m)	(2) Persons	(3) Taxation (£)
240	28·1	234,167	14,456
240– 300	9·9	36,667	5,645
301– 350	7·8	24,000	6,165
351– 400	7·9	21,067	5,973

[18] See Balopoulos, *op. cit*, 32.

Table 3·4 *Continued*

Under	(1) Income (£m)	(2) Persons	(3) Taxation (£)
401– 450	5·6	13,176	5,018
451– 500	5·7	12,000	4,603
501– 550	4·8	9,143	5,334
551– 600	4·8	8,348	4,590
601– 700	6·6	10,154	8,710
701– 800	4·8	6,400	6,754
801– 900	3·1	3,647	5,019
901–1,000	2·8	2,947	4,808
1,001–1.200	3·9	3,545	8,715
1,201–1,500	3·1	2,296	7,375
1,501 and over	3·0	1,500	9,835

The distribution of taxation by income group was the most complicated and also probably the most unreliable aspect of the study. Initially the average valuation of farms by size was calculated by means of Table 2 (Total family income per farm) and Table 3 (Family income per £ valuation). This in turn was multiplied by the number of persons whose principal occupation was returned as farming by size of farm in the 1951 Census[19] to give the total valuation by farm size. The total tax assessed under Schedule B (given in the Annual Reports of the Revenue Commissioners) was then allocated, weighted by valuation, to the various farm sizes. In order to convert farm sizes into income groups Table 9 of the National Farm Survey 'Frequency Distribution of Farms by Family Farm income (3 year averages)' was used, to give column (3) of Table 3·4. These figures were then added to Reason's figures and taxation and income per person was calculated. This was then converted into log form and the regression equation of the data was:—

$$\log PT = -7·7657 + 3·2112 \log PI \qquad (R^2 = 0·9226)$$
$$(0·2583)^\star$$

The individual marginal tax rates were calculated and when weighted by the total income received by each group a result of 0·1018 (which when corrected crudely for cross-section bias, as was done above, gives a figure of 0·120 for the aggregate marginal tax rate).

[19] *Fourth Report of the Income Tax Commission*, Pr 5731, Dublin Stationery Office, 1961, appendix III Table 1.

A second method used as a check was to split up the data into two sections putting all the income groups of under £700 p.a. into the first and the remainder into the second group. The regression of tax on income was then calculated in each case with the following results:—

Under £700
Including agriculture

$$PT = -12\cdot46 + 0\cdot0556PI \qquad (R^2 = 0\cdot9107)$$
$$(0\cdot0063)\star$$

Over £700
Including agriculture

$$PT = -249\cdot12 + 0\cdot3351PI \qquad (R^2 = 0\cdot9801)$$
$$(0\cdot0239)\star$$

The implications are that the marginal rate of tax for the under £700 group (at 1954–55 rates) can be put at 0·0556 and that for the over £700 group at 0·3351. Thus, if a rise in total income of 1 per cent is attributed to the under £700 group and the over £700 group in the same proportion as their present income, one can deduce a weighted marginal rate of 0·11 which when adjusted crudely for bias is slightly on the high side at 0·13 (a similar but more pronounced tendency to overshoot the mark was found by Prest.[20])

It must not be surmised that, because these results coincide with the earlier estimate, they prove the time-series result to be perfect. It was assumed in the cross-section approach that taxation is directly related to valuation[21]. There is also a substantial difference between the bases in the two approaches: the time-series results were based on personal income and the cross-section bases were derived from Table 3·2 of the National Income and Expenditure Accounts. Subject to these reservations the cross-section results are quite satisfactory and imply that the time series equations are reliable enough for use both in future macro-models and for forecasting purposes.

[20] A. R. Prest *op. cit.* p. 588.
[21] Thus coming up with the highly unlikely estimate of taxation on farmers with income of under £240 per annum.

7

CORPORATE INCOME TAXATION

To complete the subject of the flexibility of taxes on incomes, taxes on corporate income are studied in this section. The method adopted is similar to that for personal income taxes, the data used being for the fourteen fiscal years 1954–55 to 1967–68. The figures obtained in the previous section for income and surtax on personal incomes were deducted from the figures for income tax (including surtax) in Table A17 of the National Income and Expenditure Accounts[22] to give income tax on undistributed corporate income. The resultant figures, together with those for Corporation Profits Tax, were then adjusted by the method described earlier to give the four series at constant tax rates in Table 3·5.

TABLE 3·5

Standardised Corporate Income Tax and Corporation Profits Tax (£m)

	1954–55 Rates		1967–68 Rates	
	CIT	CPT	CIT	CPT
1954–55	7·1	2·9	7·4	6·6
1955–56	7·8	3·2	8·2	6·8
1956–57	7·7	3·1	8·1	6·7
1957–58	8·0	2·9	8·4	6·6
1958–59	6·9	2·8	7·2	6·5
1959–60	8·0	3·0	8·2	6·6
1960–61	6·7	3·3	7·1	6·8
1961–62	9·9	3·7	9·7	7·0
1962–63	11·5	4·5	10·9	7·5
1963–64	12·5	4·5	11·6	7·5
1964–65	14·7	5·0	13·2	8·4
1965–66	15·1	5·5	13·5	9·3
1966–67	13·9	5·6	12·6	9·4
1967–68	9·2	7·2	8·3	12·1
1968–69	14·8	7·6	13·4	12·8

[22] *National Income and Expenditure 1968,* Prl. 1064, Stationery Office, March 1970 and earlier issues.

An initial reservation should be set out at this stage regarding the method of adjustment of tax receipts to constant rates. This is that corporate tax receipts in one particular year include payments of tax for previous years' assessments (the same is true for personal income tax but the problem is not likely to be of the same magnitude as for companies) and in these cases the rates of tax charged and the allowances granted are governed by the law in force for the year of assessment concerned. Therefore this is a composite marginal rate of tax for receipts in a particular period and not a true marginal tax rate based on accruals in a given year. In order to have some idea as to the relationship of tax yield to profit earned in particular periods and to try to estimate the relative weights of particular years an analysis was made of the magnitude of profits falling within the different quarters of the year. The basic data used was that contained in *Trade Union Information*[23] which publishes the main figures in the balance sheets of public companies reporting in the year. It was decided that rather than use the percentage distribution of accounting periods throughout the year (the average figures calculated for 1966 and 1967 were 24 per cent in Quarter 1, 16 per cent in Quarter 2, 17 per cent in Quarter 3, 43 per cent in Quarter 4 and are broadly similar to those obtained by McDowell[24] from the Revenue Commissioners for 1964—31·5 per cent in Quarter 1, 15·3 per cent on Quarter 2, 15·3 per cent in Quarter 3 and 37·8 per cent in Quarter 4) it would be more appropriate to obtain a distribution of reported profits—as one or two large concerns could substantially influence tax payment patterns (the hundred-plus firms analysed representing nearly £20 million of pre-tax profits in 1967). The weights calculated for profits falling into various quarters were as follows (most of the accounting periods ending at end-quarter dates).

Quarter 1	19·5
Quarter 2	19·5
Quarter 3	22·7
Quarter 4	39·3

[23] *Trade Union Information* (published monthly by the Irish Congress of Trade Unions), various issues 1967, 1968.
[24] M. McDowell *The Irish Tax Structure: Income Elasticity of Yield and Built-in Flexibility,* unpublished M.A. Thesis, National University of Ireland.

The C.S.O. figures for profits included in the national accounts are for profits of firms with accounting dates falling between 1 July of the year in question and 30 June of the following year. Assuming profits are spread evenly over the firm's year and that all firms report at the end of the quarter to which they are apportioned, then approximately 5 per cent of the C.S.O. figures for profits in year t will represent profits actually earned in year $t-1$, 80 per cent in year t and 15 per cent in year $t+1$. The C.S.O. figures for profits also treat losses in a different way than do the Revenue Commissioners. The C.S.O. deduct losses to arrive at their total profits figures: the Revenue Commissioners treat losses as being equal to zero profits.

The administrative procedures for corporation profits tax produce an eight month lag between the end of a firm's accounting period and the collection of the tax. CPT receipts in the fiscal year $t/t+1$ represent 5 per cent profits actually earned in year $t-2$, 80 per cent in year $t-1$ and 15 per cent in year t, thus indicating a one year lag between the figures for receipts of tax and the C.S.O. published yearly profit figures (if all the tax was actually paid after 8 months, but see Table 82 in the 44th Report of the Revenue Commissioners for a statement of the assessments outstanding for more than two years).

Income tax on companies is slightly more complicated and the spread of time periods can be even more diverse than CPT. Tax is assessed on the basis of the income of the previous accounting period and is paid on 1 January. An example of the importance of the end of the accounting period date can be given as follows. A company whose accounts are made up to 30 April annually pays income tax for 1970–71 on its income for the year ended 30 April 1969 and the tax is payable on 1 January 1971, i.e. twenty months after the end of the accounting period—whereas if accounts are made up to 31 March the tax due on 1 January 1971 is that based on the accounts for the year ended 31 March 1970—i.e. the tax is due 9 months after the end of the accounting period[25]. Taking account of the ends of accounting periods therefore as before, 15 per cent of the tax paid in the financial year $t/t+1$ relates to profits earned in year $t-2$, 80 per cent to year $t-1$ and 5 per cent to year t, thus indicating a lag of between one and two years.

[25] *Sixth Report of the Commission on Income Taxation*, Pr. 6030, Dublin Office, 1961, 56.

Having thus established the time period involved, it is as well to set out the other difficulties which arise when formulating corporate tax equations. First, corporation profits tax and income tax paid by companies must be separately distinguished not merely because of the difference in the time period involved, but also because the base of corporation profits tax is total profits and that for corporate income tax is undistributed profits (account has already been taken, in the figures for the marginal rate of taxation on persons, of tax paid on dividends). Secondly, corporation profits tax, prior to 1961–62 and from 1966–67, until the supplementary budget of 1970 was allowed as a deductable expense in computing business profits for standard income tax purposes and therefore should be deducted from the base when computing the marginal rate of corporate income tax. Thirdly, as mentioned before, losses are included as a minus figure in the tax base whereas they should be treated as zero and in addition to this no account is taken of losses carried forward.

Various equations were fitted by ordinary least-squares using (1) corporation profits tax, (2) corporate income tax and (3) both combined as the dependent variables and various definitions of corporate profits as the independent variables. The equations were nowhere near as conclusive as those found for personal income tax.

The main disturbing feature which emerged was that, in most of the equations containing current and lagged income coefficients, the coefficients varied as between positive and negative values, this in turn probably accounting for the poor DW values. Various adjustments were made to the equations to try and eliminate the negative coefficients but no significant improvement was brought about and its analysis must await another occasion. In all the equations the time variable was insignificant but there did appear to be some evidence of a negative time trend in the corporation tax equations which is probably due to the increasing importance of tax relief on exports. One would have expected that the export profits tax relief would have exerted a stronger pull through the time trend on both corporation and income tax but its effect is likely to have been incorporated in the income coefficient as the difference between the marginal rate of tax at 1954–55 rates and 1967–68 rates, despite no large decrease in income taxation rates, would seem to indicate.

The first difference equations proved to be the most satisfactory estimators of the corporate income tax function from a statistical point of view.

CORPORATE INCOME TAX

1955–56 RATES

(14) $\Delta CIT_t = -0.319079 + 0.33467\Delta UP_{t-1} - 0.05834\Delta UP_{t-2}$
$\qquad\qquad\quad (0.06765)^\star \qquad\qquad (0.09948)$
$\qquad R^2 = 0.7306 \quad SE = 1.348 \quad DW = 1.213$

(15) $\Delta CIT_t = -0.013199$
$\qquad +0.29349\Delta (UP - CPT)_{t-1} + 0.01357\Delta (UP - CPT)_{t-2}$
$\qquad (0.04438)^\star \qquad\qquad\qquad (0.04439)$
$\qquad R^2 = 0.8003 \quad SE = 1.155 \quad DW = 1.938$

(16) $\Delta CIT_t = -0.094553$
$\qquad -0.07855\Delta UP_t + 0.33091\Delta UP_{t-1} - 0.03629\Delta UP_{t-2}$
$\qquad (0.06318) \qquad\quad (0.06610)^\star \qquad\quad (0.09870)$
$\qquad R^2 = 0.7667 \quad SE = 1.315 \quad DW = 1.501$

1967–68 RATES

(17) $\Delta CIT_t = -0.314347$
$\qquad +0.29690 \, \Delta UP_{t-1} - 0.06870 \, \Delta UP_{t-2}$
$\qquad (0.05749)^\star \qquad\quad (0.08453)$
$\qquad R^2 = 0.7533 \quad SE = 1.145 \quad DW = 1.323$

(18) $\Delta CIT_t = -0.029632$
$\qquad +0.26091\Delta (UP - CPT)_{t-1} + 0.00620 \, \Delta (UP - CPT)_{t-2}$
$\qquad (0.03881)^\star \qquad\qquad\qquad (0.03882)$
$\qquad R^2 = 0.8077 \quad SE = 1.011 \quad DW = 1.815$

(19) $\Delta CIT_t = -0.10409$
$\qquad -0.07356\Delta \, UP_t + 0.29338\Delta \, UP_{t-1} - 0.04805 \, \Delta UP_{t-2}$
$\qquad (0.05280) \qquad\quad (0.05524)^\star \qquad\quad (0.08248)$
$\qquad R^2 = 0.7934 \quad SE = 1.099 \quad DW = 1.661$

Where CIT = corporate income taxation
$\qquad\quad UP$ = undistributed profits of companies before tax
$\quad UP\text{–}CPT$ = undistributed profits of companies after corporation profits tax
$\qquad\quad CPT$ = corporation profits tax

The forecasting power of these equations was very precise for 1968–69 in contrast to the poor performance of equations with the raw data. Inspection of the residuals in the original corporate income tax equations showed that for the last two observations there was an excess of the estimated tax yield over the actual yield. A similar problem in the other direction arose in Prest's[26] equations for personal income tax and the procedure used here was to add the excess of the actual over the projected figure for the previous year to the regression projection. As Prest admits, there is another way in which the correction factor could be calculated and that is by averaging the excess of the actual over the estimated yield for the two year period. It was considered therefore that this adjustment procedure was slightly crude and that in any case there was no need for it as the first difference equations, which gave accurate forecasts, could be taken as more representative of the corporate tax function. The equations suggest that the long-term marginal corporate income tax rate with respect to undistributed profits is between 0·22 and 0·28 at 1954–55 rates and between 0·17 and 0·22 at 1967–68 rates with a preference for the lower end of the range because of the smaller standard errors. If it is desired to use the base adjusted for payments of corporation profits tax the marginal rate would be around 0·30 at 1954–55 rates and 0·26 at 1967–68 rates.

The corporation profits tax equations produced the same difficulties as can be seen from the following equations.

CORPORATION PROFITS TAX

1954–55 RATES

(20) $CPT_t = -0.921408$
$+0.06732 TP_t - 0.03949 TP_{t-1} + 0.09150 TP_{t-2} - 0.17953 T$
$(0.01975)^\star \quad (0.02490) \quad (0.02180)^\star \quad (0.06525)^\star$
$R^2 = 0.9632 \quad SE = 0.228 \quad DW = 1.471$

(20a) $\Delta CPT_t = -0.125220$
$+0.05854 \Delta TP_t - 0.01628 \Delta TP_{t-1} + 0.07184 \Delta TP_{t-2}$
$(0.01841)^\star \quad (0.02228) \quad (0.02797)^\star$
$R^2 = 0.6343 \quad SE = 0.316 \quad DW = 2.945$

---

[26] Prest *op. cit.* p. 582.

1967–68 RATES

(21) $CPT_t = 1 \cdot 862932$
$+ 0 \cdot 09785\,TP_t - 0 \cdot 13173\,TP_{t-1} + 0 \cdot 18473\,TP_{t-2} - 0 \cdot 26932\,T$
$(0 \cdot 04599) \qquad (0 \cdot 05798) \qquad\quad (0 \cdot 05075)^\star \qquad\quad (0 \cdot 15193)$
$R^2 = 0 \cdot 9245 \quad SE = 0 \cdot 530 \quad DW = 1 \cdot 653$

(21a) $\Delta\,CPT_t = 0 \cdot 159594$
$+ 0 \cdot 10300\Delta\,TP_t - 0 \cdot 04704\Delta\,TP_{t-1} + 0 \cdot 08500\Delta\,TP_{t-2}$
$(0 \cdot 03127)^\star \qquad\quad (0 \cdot 03784) \qquad\qquad (0 \cdot 04751)$
$R^2 = 0 \cdot 5924 \quad SE = 0 \cdot 537 \quad DW = 2 \cdot 136$

Where CPT = corporation profits tax
$\qquad\ TP$ = trading profits before tax
$\qquad\ \ T$ = time.

The long-term marginal corporation profits tax rates with respect to trading profits of companies can be put at about 0·11–0·12 at 1954–55 rates and 0·14–0·15 at 1967–68 rates.

Additional regressions were run combining corporate profits tax and corporate income tax at 1967–68 rates and expressing the total as a function of undistributed profits of companies before tax. The statistical results improved somewhat, the following being the best results.

CORPORATE INCOME TAX PLUS CORPORATION PROFITS
TAX: 1967–68 RATES

(22) $(CIT + CPT)_t = 7 \cdot 5480 + 0 \cdot 303733\,UP_{t-1}$
$\qquad\qquad\qquad\qquad\qquad\quad (0 \cdot 020445)^\star$
$\qquad R^2 = 0 \cdot 9484 \quad SE = 0 \cdot 785 \quad DW = 1 \cdot 388$

(22a) $\Delta(CIT + CPT)_t = 0 \cdot 01086 + 0 \cdot 29821\Delta\,UP_{t-1}$
$\qquad\qquad\qquad\qquad\qquad\qquad\quad (0 \cdot 044362)^\star$
$\qquad R^2 = 0 \cdot 7901 \quad SE = 0 \cdot 943 \quad DW = 2 \cdot 360$

(23) $(CIT + CPT)_t = 7 \cdot 53949 - 0 \cdot 00827\,UP_t + 0 \cdot 31301\,UP_{t-1}$
$\qquad\qquad\qquad\qquad\qquad\quad (0 \cdot 03992) \qquad\quad (0 \cdot 04960)^\star$
$\qquad R^2 = 0 \cdot 9486 \quad SE = 0 \cdot 819 \quad DW = 1 \cdot 443$

(23a) $\Delta(CIT + CPT)_t =$
$\qquad -0 \cdot 046359 + 0 \cdot 01634\Delta\,UP_t + 0 \cdot 30001\Delta\,UP_{t-1}$
$\qquad\qquad (0 \cdot 04585) \qquad\quad (0 \cdot 04535)^\star$
$\qquad R^2 = 0 \cdot 7925 \quad SE = 0 \cdot 970 \quad DW = 2 \cdot 32$

These equations suggest that the combined long-term marginal rate of corporate taxation with respect to undistributed profits is approximately 0·30 which is comparatively near the total obtained from summing the two separate calculations.

Unfortunately, it is not possible to check these results by cross-section analysis as there are no statistics available on the breakdown of profits and tax paid by companies operating within the State or a representative sample thereof—any published statistics usually relate to Irish public companies which would tend to produce a strong upward bias in the estimate of the rates, excluding, as this sample would, foreign firms operating here and exporting a large proportion of their output.

CONCLUSION

The only other work done on the flexibility of Irish taxation came to the author's attention towards the end of the preparation of this study. McDowell's[27] approach to the problem differs substantially from ours in that in the case of income tax he analyses the elasticity of taxes (using the correct but rather complicated method of relating yield to taxable income and taxable income to the appropriate national accounts definition, in contrast to the present method of using national accounts definitions together with constant tax rates and neglecting the relationship of the legislated tax base to the appropriate national accounts definition) under the various schedules as opposed to the method used here of distinguishing between taxes paid by persons and taxes paid by companies.

He also studies customs and excise taxes and turnover tax and quantifies the built-in stability imparted by taxation, showing that about 35 per cent of fluctuations in income are prevented by the flexibility of non-PAYE Income Tax, Surtax and Corporation Profits Tax. However, if an import function in line with that found by McAleese[28] for Ireland were incorporated into this model, the percentage of fluctuations reduced by the flexibility of these taxes is brought down to 11 per cent.

This analysis provides a basic building block for future studies on this topic. Lines of enquiry might be extended to analyse the

[27] M. McDowell, *op. cit.*
[28] D. McAleese, *A Study of Demand Elasticities for Irish Imports,* Dublin, Economic and Social Research Institute, Paper 53, 1970.

effects of these and other taxes and transfers on stability, to integrate these results into a dynamic model, to assess the relative importance of stabilisers in expansions and recessions and to study the dampening effect of built-in flexibility on inflationary pressures.

APPENDIX
ADJUSTMENT OF PERSONAL INCOME DATA
TO CORRESPOND TO THE FINANCIAL YEAR

In order to derive estimates of personal income for the financial year a series for quarterly personal income was required. Initially annual disposable income figures were taken and transformed into a quarterly series. The method used was that set out by McAleese[29]. By this method of calculation annual observations of disposable income figures are taken and a straight line interpolation is made between them assuming that the annual observations lie at the end of the second quarter of each year. From this, values for the last two quarters of the first year and the first two quarters of the second year can be calculated. Quarterly changes in the output of transportable goods industries are then superimposed on the trend of the calculated values making allowance for the varying share of transportable goods in total output. This produces estimates of quarterly personal disposable income for the period 1953-69.[30]

The next step was to arrive at personal income figures and for this estimates were required for taxes on personal income. The method used for calculating taxes on personal income was as follows. Income Tax (including surtax) receipts were ascertained quarterly from 'Receipts into and issues out of the Exchequer'[31] which were seasonally adjusted[32] to allow for the fact that pay-

[29] D. McAleese, *op. cit.*

[30] Mr McAleese was kind enough to provide me with his calculations of quarterly figures at annual rates on constant prices for 1956–1965.

[31] *Iris Oifigiúil,* various publications, and Central Bank Quarterly Bulletins.

[32] Initially the adjustment method used was that set out in C. E. V. Leser, *Seasonality in Irish Economic Statistics,* Economic Research Institute, Paper 26, 1965 —the correction factor for the latest year being derived from data for the preceding five year period, but this was found to be unsuitable particularly in recent years because of the strong underlying trends in the quarterly figures. The method used was basically a compromise. The above method was used but the correction factors were applied to the year in the middle of the five year period instead of to the year following the period, thus eliminating the trend. The factors for years after 1966 were derived by extrapolating the previous four correction factors.

ments in the last quarter of the financial year were abnormally high—due principally to corporation tax payments. The seasonally adjusted quarterly figures for total tax payments were then expressed as a percentage of total tax paid during the year and these percentages were then applied to the figures in the National Accounts for taxes on personal income and quarterly figures were arrived at. These tax payments were then added to the calculations of disposable income to get quarterly figures for personal income which were added together to give the figures for the financial years.

CHAPTER IV

Hidden Subsidies in Irish Electricity Supply[1]

J. A. BRISTOW

THIS essay is concerned with the measurement of the subsidies received or paid by consumers of electricity in Ireland. But before this can be done, two logically prior questions must be investigated. First, how are subsidies to be defined in this case? and second, do they exist? The first section looks at these questions; subsequent sections deal with the estimates; and the final section discusses the implications for resource allocation and income distribution. Policy prescriptions are eschewed: some alternatives are mentioned, but it is left to the reader to draw conclusions concerning the desirability or otherwise of a change in policy.

THE NATURE OF HIDDEN SUBSIDIES

A subsidy is an economically unrequited transfer of resources— a gift—which may be in money or in kind. In national income accounting they are treated as transfer payments, and when given by the government to the private sector they are regarded as taxes but with the opposite sign. (There may be argument about whether any given payment is properly treated in this way—e.g.,whether motor vehicle duty should be regarded as a tax or as a payment for the use of roads, or whether unemployment benefit is a transfer or the *quid pro quo* for national insurance contributions—but the principle is clear enough.) The theory of public finance treats government subsidies as negative taxes at both the macroeconomic and microeconomic levels. Thus,

---

[1] The author is indebted to several people in the relevant bodies—notably Mr J. B. O'Donoghue of the Electricity Supply Board and Mr C. F. Fell of Bord na Móna. The usual disclaimer of responsibility applies not only to them but also to their employers.

macroeconomically they are an injection into the private income flow as taxes are a withdrawal. Microeconomically, they may influence costs, prices, resource allocation and income distribution in a way symmetrical to the influence of taxes. Since this is an exercise in microeconomics, the latter will be given more attention. The discussion will be conducted in terms of cash payments by the government, although, as will become apparent, all that is said is applicable *mutatis mutandis* to other forms of subsidisation.

Subsidies can be categorised according to who initially receives them (although for certain purposes other taxonomies are more appropriate). Thus, a subsidy may be paid to (a) a person or organisation defined by name, (b) persons defined by income, (c) persons living in certain regions, (d) consumers of a particular product, (e) producers of a particular product, (f) producers in certain regions, (g) users or suppliers of particular factors of production, etc. These are likely to differ in their effects upon resource allocation and income distribution. Perhaps the most useful distinction for present purposes is between those subsidies which achieve their effects through their impact on relative prices of goods and services and those which act in a different way. Thus, all subsidies are likely to change the distribution of income but some do this directly and some indirectly in a way dependent upon cost functions, business behaviour, the utility functions of consumers, etc. Similarly, a subsidy may influence the allocation of resources directly by changing price ratios or relative factor rewards or indirectly by causing discriminatory shifts in demand. Therefore, those subsidies designed primarily for reallocating resources will usually achieve their effect by impacting directly on relative prices, whereas those designed to redistribute income will influence relative prices only as a secondary effect.

As a preface to the discussion of the particular subsidies under review, it is worth setting out more specifically how these effects come about. To illustrate the argument subsidies to producers of particular goods and those to persons in particular income groups are taken. The first type is typically specific—so much per unit of output. This reduces average and marginal costs and, if producers are profit maximisers (with one exception) or mark-up pricers, will lead to a reduction in market price. The exception is the pure monopolist who will take the subsidy as an addition to profit. The same may occur if there is collusive oligopoly or some other form of non-marginalist behaviour. But the result may be the same in the long run since this addition to profit may attract capital into this industry, the expansion producing a reduction in prices. If the price reduction does not take place, this producer subsidy has a redistributive effect in favour of capital.

If the price reduction occurs—and the usual immediate purpose of this type of subsidy is to achieve such a reduction—the impact upon the allocation of resources will depend upon the form of consumers' utility functions. If there are close substitutes, there will be a large substitution effect and demand will switch significantly to the subsidised good. Even if the marginal rate of substitution of this for other commodities is low, or in the extreme case if utility functions are discontinuous at the pre-subsidy level of consumption (which may be so for electricity in the short run), the whole of the impact will be in the form of an income effect which would increase saving and/or the demand for other goods.

Whichever happens, there is a redistributive effect: welfare has been transferred to those consuming any of this good in the pre-or post-subsidy situation at the expense of those providing the subsidy. Clearly, the net redistributive effect depends upon how one disaggregates the community and upon how the subsidy is financed, Thus, it is conceivable that people benefit by the subsidy to exactly the same extent to which they pay taxes to finance it, but this would be very rare. More usually there would be a net redistribution because the beneficiaries are totally distinct from the payers or because the beneficiaries are only part of the paying group.

In a comparable way, the subsidy designed primarily as a redistributive device may have allocative effects. This is because the marginal propensity to consume an individual good may vary as income varies. Thus, if the marginal propensity to consume a good rises as income rises, a subsidy which transfers income from the rich to the poor will reduce total demand for that good. Conversely, if the marginal propensity to consume falls as income rises, such a subsidy will increase demand.

Then, the distributive effects of both types of subsidy may influence the proportions in which factors are combined. A subsidy which, directly or indirectly, transfers income from capital-owners to labour will raise the supply price of capital and lower that of labour (except where the supply curves of either are 'backward-bending').

One final point on subsidies in general. Clearly, subsidies have to be paid for—i.e. somewhere in the system there is a tax (or its equivalent) which would be reduced if the subsidy did not exist, or there is an item of public expenditure which would be increased in such circumstances. Since earmarked taxes are rare, it is usually not possible to range against a subsidy the tax which provides the funds, although, as will be seen, this is possible for the subsidies discussed in this essay. If it were possible, one could adopt the ideal procedure of looking at the net allocative and distributive effects of the tax and subsidy together. This is especially important in normative analysis. For instance, in partial equilibrium terms a subsidy may appear desirable—because it benefits the right people and/or moves the system towards an optimal allocation—but a more general analysis may show that the tax has opposite and greater effects and so the community would be better off without the subsidy. Subsidies financed from a general fund can usually only be investigated in partial terms, but such a procedure is less than ideal.

The discussion so far has been of overt grants. The subject under review is hidden subsidies, but the foregoing was necessary because such subsidies are here defined as policies which have effects similar in kind to overt subsidies. Thus, a hidden subsidy is one whose primary objectives could in principle be achieved by a cash grant. Of course, to qualify as a hidden subsidy under this definition it is not necessary that all effects be similar to those of the direct grant. An example from the private sector will illustrate this. The statutory requirement that private motorists purchase third-party insurance is, under our definition, a hidden subsidy to those in the population who may suffer damage and cannot obtain satisfaction because the motorist has insufficient assets. The same primary result could be achieved if the government paid everyone a grant to buy their own insurance against damage from motorists. If the grant comes from general taxation, and under the present system the subsidy is paid by the motorist, the distributive and secondary allocative effects will be different, but the primary effect is the same.

This illustration is particularly apt since in the cases to be investigated here the subsidies arise from formal or informal governmental decree and involve one definable section of the population benefitting at the expense of another definable section.

Having therefore defined the subject and briefly set out the terms in which it will be discussed, the next task is to show that the subject actually exists. This can be done most easily by simply listing the subsidies to be dealt with.

First, since the war, the Electricity Supply Board has been required to build turf-fired generating capacity (and one station fired by native coal). The first was commissioned in 1950 and the last of such sets is currently being installed. The subsidy arises because unit fuel costs and capital costs per unit of capacity are higher than those at oil-fired stations. In the early 1950s, when coal and oil prices were high, turf was the least-cost fuel. It could, therefore, be argued that electricity consumers are not subsidising producers of native fuel but are merely suffering from an unforeseen change in relative fuel prices. But this argument is not sustainable. With the exception of the immediate post-Suez years when oil prices were abnormally high, unit generating costs of native fuel have always been higher than those of imported fuels[2], whereas most of the turf-fired capacity has been installed since that date. The E.S.B. have consistently argued that the use of turf was not in the interests of electricity consumers, and one is justified in accepting this policy as one imposed upon the Board by the government. It qualifies as a hidden subsidy under the definition used here because the same primary objective—a high demand for turf by the E.S.B.—could be achieved by a grant either to Bord na Móna (the public enterprise supplying practically all the turf used by the E.S.B.) so that they could supply turf for generating purposes at a price competitive with that of oil or to the E.S.B. so that the excess costs of native-fuel utilisation could be borne without influencing average electricity prices.

Second, there is a system of cross-subsidisation within the E.S.B. Because of the scattered nature of the load, costs of supply to rural consumers are higher than those to urban consumers. There are separate tariffs for the two groups of consumer, but

---

[2] The first peat-fired sets were commissioned in 1950. In terms of price per Btu, only in 1957 and 1958 was sod peat cheaper than coal or oil, and only during the years 1957–59 was milled peat cheaper. See J. L. Booth, *Fuel and Power in Ireland: Part II Electricity and Turf,* Economic Research Institute, Paper No. 34, Dublin 1966.

the government does not permit the difference in charges to equal the difference in costs. The government makes a contribution to the costs of rural connection, but this is insufficient to cover even the extra connection costs and no contribution at all is made with respect to the extra current costs. Rural operations therefore run a deficit which is financed from a profit on urban account. Thus, if the Board's financial objective were applied separately to the two operations urban charges would be lower and rural charges higher: that is the subsidy. Again, the same immediate objective of stimulating rural consumption could be achieved by a grant either to rural consumers or to the E.S.B. to cover the cost difference.

Third, the E.S.B. receive a substantial subsidy in the form of exemption from rates on generating stations, transformer stations and transmission lines. This is the equivalent of grants from the local authorities in whose areas the rate-relieved property is located, and could be replaced by an equivalent sum from the Exchequer.

All these subsidies clearly exist. But there is another form of subsidisation which must be mentioned but whose existence is a matter of controversy. This subsidy would occur if the Board's general pricing policy were different from what it should be, and the controversy is over what is the desirable pricing policy. There are a large number of candidates, but to keep the discussion within bounds we limit ourselves to:

(a) Marginal cost pricing with no financial constraint—i.e. the deficits arising from the application of marginal cost pricing in this industry are accepted and made good by the Exchequer.

(b) Marginal cost pricing with the constraint that deficits are impermissable. This would require the tariff to include a part not related to marginal cost.[3]

(c) Profit maximisation.

(d) Average cost pricing, with cost including a 'normal profit' component. The proper level of normal profit is a matter for argument and so there can be a number of possibilities under this heading. The same problem arises with (b) because the

[3] The bulk supply tariff of the CEGB is constrained in this way—see R. L. Meek, 'The New Bulk Supply Tariff for Electricity', *Economic Journal,* LXXVIII, (March 1968).

8

costs to be covered have to be known before it can be known whether or not a deficient exists.

The tariff of the Electricity Supply Board is multi-part and is basically of form (d), with some concessions in the direction of (b). This being the case, one has the following possibilities. If one believed in (a), electricity prices are at present too high, and so electricity consumers are subsidising the rest of society. If one believed in (b), prices are on average correct (assuming the financial constraint is defined to one's satisfaction), although certain consumers will be paying above, and others below, marginal cost and some cross-subsidisation exists within the industry. If (c) is one's optimum, prices are too low, and the users of electricity are being subsidised.

The author is not a pure marginal-cost pricer, for the usual second-best reasons. He would accept as sensible, however, tariffs which distinguish between peak and base load, as long as the distinctions are not made so fine as to create significant pricing costs, and as long as each part of the tariff is designed to cover all the relevant costs. But this is not marginal-cost pricing: it consists of dividing the market into sectors and, as far as practicable, applying average-cost pricing separately to each sector. An element of subsidisation would then arise if a sector were not covering its costs—e.g. if consumers at the peak were not paying the full costs of meeting peak load. This possibility, which very definitely exists for the E.S.B., is left for another study.

Profit maximisation is obviously a non-starter. In the short-run anyway, demand must be almost completely price-inelastic, and profits could be enormous. Since this monopoly position was reached with state assistance, it would be hard to justify such exploitation of this position.

If (d) is, in general, accepted as the correct policy, the remaining possibility is that the level of 'normal profit', which here is not determined competitively, is inappropriate. If too low, prices are too low and *vice-versa*. Unlike British nationalised industries[4], the financial objective of the E.S.B. is not defined as a target rate of return. Under the Electricity Supply Act of 1927, the Board must break even after certain, defined costs have been covered.

[4] See *The Financial and Economic Obligations of the Nationalised Industries*, Cmnd. 1337, London, H.M.S.O. 1961 and *Nationalised Industries: A Review of Economic and Financial Objectives*, Cmnd. 3437, H.M.S.O., London 1967.

The Board's form of accounting reflects this and each year the amount recorded as net surplus (occasionally deficit) is trivial (recently around one quarter per cent of capital). Two points can be made here. First, despite appearances, the financial objective *requires* the earning of what in normal usage would be termed profit. This is because the Board charges to revenue not only depreciation, but also a sinking fund provision. This rather misleading form of accounting, which charges to revenue an item normally counted as part of profit, seems to exist because the 1927 Act lists capital repayment provisions as one of the costs to be covered. It is not even possible, from the published accounts, to extract the value of this profit, since it is aggregated with interest payments. This is especially regrettable since otherwise the E.S.B.'s accounts are probably the most informative published by an Irish public enterprise.

Thus, some profit is earned. But, if one knew what that profit was, one could still argue about whether it were too large or too small. On the one hand one might suggest that, for the E.S.B., *any* positive profit is too large. There is no equity capital, and profit would be a pure rent arising from the monopolistic position of the Board. On the other hand, one could claim that profit is low if it is lower than that earned by the producers of competing goods (i.e. if profits are too low it indicates that prices are too low, thus distorting consumer choice between the competing goods). When one considers the utter confusion and lack of economic rationale of the methods of deriving target rates of return for nationalised industries in Britain, the author feels justified in avoiding further incursions into this jungle.

The result of this is that there is here no imputation of subsidies to the general level of the Board's tariff. The point of reference will be the current financial objective. The subsidies to be measured all arise because, given that objective, government policies require electricity sales revenue (and, therefore, average electricity prices) to be different from those required in the absence of such policies.

THE USE OF NATIVE FUEL

The E.S.B. use three kinds of native fuel: sod and milled peat supplied by Bord na Móna, hand-won sod peat supplied by private producers, and coal. Of these, the last two are insignificant: the hand-won turf serves four 5MW stations in the West whose output in 1968–69 was only 39·4 million Kwh, and the coal serves a 15MW station in Co. Roscommon whose output in that year was a mere 60·3 million Kwh. Although these are identified in the calculations, the discussion is solely in terms of the subsidy which electricity consumers pay to Bord na Móna.

The subsidy arises because government has required the E.S.B. to follow other than a cost-minimisation policy with respect to both operational and investment decisions. Thus, turf-fired capacity has been installed which has higher operating and capital costs than alternative forms of capacity, and the turf stations have been given preference in use—i.e. with a higher load factor than merit order, cost-minimising, procedures would permit.

In any year, the value of the subsidy equals the excess of actual costs over the level of costs which would have been incurred if cost-minimising investment decisions had been made in the past and if day-to-day operations were cost-minimising. (Because the financial objective is to break even exactly, excess costs can be translated directly into excess revenue requirements and therefore into excess prices.) One therefore needs to make some assumption as to what capacity would have been installed if the constraint had not existed. First, it is assumed that thermal capacity would have been installed since all the technically and economically feasible hydro developments have in fact been made. Then, the thermal capacity would have been coal or oil-fired (or more precisely, judging from the E.S.B.'s actual programme of non-turf thermal capacity since 1950, it would have been initially designed to use coal or oil if installed in the 1950s, with later use being solely of oil, and if installed in the 1960s it would have been simply oil-fired). In the years to which the estimates refer, all the Board's oil-fired capacity had been installed during the period when turf-fired stations were also coming into commission. One can therefore assume that, in the absence of the constraint, the E.S.B. would have installed oil-fired stations with the same cost characteristics as the oil-fired stations actually in operation at the time.[5] The difference between costs at turf stations and costs at the oil stations (weighted average of the three stations) is then the

[5] The relevant oil stations are Ringsend (Dublin, 270MW in 1968–69), Marina (Cork, 120MW in 1968–69) and the North Wall peaking station (Dublin, 48MW in 1968–69).

excess cost. This procedure may slightly underestimate the excess cost because there were probably unused economies of scale in generation from oil, which could have been realised if the alternative policy had been followed. But this possibility is ignored in the estimates.

Fuel costs are taken, for the purposes of this study, as the sole variable generating cost. This is because an unknown proportion of the costs listed as 'works cost' in the E.S.B.'s accounts must be overheads—e.g. certain labour must be employed if a station is to remain in commission, even if it is not used. In any case, fuel costs represent, for all thermal stations, over 75 per cent of 'works cost' and non-fuel unit works cost varies much less than fuel cost between different types of thermal station. Thus the difference in non-fuel works cost per Kwh sent out between the relevant oil stations and the B.N.M. supplied turf stations was in 1968–69 only 0·036d whereas the equivalent difference in fuel cost was 0·378d.

Excess fuel costs arise for two reasons: first, the price of turf per Btu is higher than oil, and second, turf stations are less efficient in that one ton of oil equivalent produces less electricity there than at oil stations. The explanation for the latter would seem to be with two factors—boiler pressures are lower at turf stations, and the average set and station is smaller in the turf-fired sector (the average-size set fired by B.N.M. turf is 22·8MW, compared with 36·5MW at the three relevant oil stations or 43·3MW if the older and very small North Wall station is excluded), thus sacrificing economies of scale. For 1968–69, 28 per cent of the excess fuel costs arising at stations using Bord na Móna turf could be traced to the difference in efficiency, the remainder resulting from the price difference.[6]

Figures on relative fuel costs are given in Table 4·1. In the final column is shown what total fuel costs would have been if the unit fuel cost at oil stations had ruled at native-fuel stations. If this is subtracted from actual fuel costs, one obtains the estimate of excess fuel costs shown in Table 4·2. The latter also shows these as a percentage of total revenue and so shows the percentage by which prices would be lower (without affecting profitability) if the native-fuel stations had not been built and used preferentially.

[6] For the method of calculating this figure see J. A. Bristow and C. F. Fell, *Bord na Móna: A Cost-Benefit Study*, Dublin, 1971.

TABLE 4·1

Fuel Costs

		Mn. Kwh. sent out	Fuel cost per Kwh (d)	Total fuel cost (£'000)	Fuel cost at oil cost (£'000)
Oil	1966–67	1805·5	0·310	2329·9	—
	1967–68	1983·7	0·341	2820·0	—
	1968–69	1438·9	0·429	2573·9	—
B.N.M. turf	1966–67	848·8	0·807	2855·2	1096·4
	1967–68	1055·4	0·778	3420·1	1499·5
	1968–69	1613·2	0·797	5355·5	2883·6
Hand-won turf	1966–67	27·2	1·137	128·8	35·1
	1967–68	22·4	1·344	125·4	31·8
	1968–69	39·4	1·122	184·3	70·4
Native coal	1966–67	72·5	0·738	223·0	93·6
	1967–68	72·5	0·734	221·7	103·0
	1968–69	60·3	0·780	195·9	107·8
Total native fuel	1966–67	948·5	0·811	3207·0	1225·1
	1967–68	1150·3	0·786	3767·3	1634·4
	1968–69	1712·9	0·804	5735·7	3061·8

TABLE 4·2

Excess Fuel Costs

		Excess fuel costs (£'000)	As % of revenue
B.N.M. turf	1966–67	1758·8	6·6
	1967–68	1920·6	6·4
	1968–69	2471·9	7·2
Hand-won turf	1966–67	93·7	0·4
	1967–68	93·6	0·3
	1968–69	113·8	0·3
Native coal	1966–67	129·4	0·5
	1967–68	118·7	0·4
	1968–69	88·1	0·2
Total native fuel	1966–67	1981·8	7·4
	1967–68	2132·9	7·1
	1968–69	2673·9	7·8

A similar treatment is given to capital costs in Tables 4.3 and 4.4. Capital costs here are the sum of depreciation, interest and sinking fund (i.e. profit) and so exceed what is normally defined as capital costs. Thus, they vary slightly from year to year for a given capacity, whereas they should be constant. But, from the Board's accounts it is not possible to infer true capital costs and there was no alternative but to adopt this procedure. However, the error is slight.

TABLE 4·3

Capital Costs

		Capacity (MW)	Cap. cost per MW (£)	Total cap. cost (£'000)	Capacity in alt. sit. (MW)	Cap. cost in alt. sit. (£'000)
Oil	1966–67	438	3621·5	1586·2	—	—
	1967–68	438	3816·7	1671·7	—	—
	1968–69	438	3884·3	1701·3	—	—
B.N.M. turf	1966–67	347·5	4747·2	1649·7	332	1202·4
	1967–68	387·5	4907·4	1901·6	370	1412·2
	1968–69	387·5	5010·3	1941·5	370	1437·2
Hand won	1966–67	20	5735·1	114·7	Nil	Nil
turf	1967–68	20	5826·4	116·5	Nil	Nil
	1968–69	20	5930·5	118·6	Nil	Nil
Native coal	1966–67	15	5022·9	75·3	Nil	Nil
	1967–68	15	5140·3	77·1	Nil	Nil
	1968–69	15	5266·5	79·0	Nil	Nil
Total native	1966–67	382·5	4809·7	1839·7	332	1202·3
fuel	1967–68	422·5	4959·2	2095·2	370	1412·2
	1968–69	422·5	5062·9	2139·1	370	1437·2

TABLE 4·4

Excess Capital Costs

		Excess cap. costs (£'000)	As % of revenue
B.N.M. turf	1966–67	447·4	1·7
	1967–68	489·4	1·6
	1968–69	504·3	1·5
Hand-won turf	1966–67	114·7	0·4
	1967–68	116·5	0·4
	1968–69	118·6	0·3
Native coal	1966–67	75·3	0·3
	1967–68	77·1	0·3
	1968–69	79·0	0·2
Total native fuel	1966–67	637·4	2·4
	1967–68	683·0	2·3
	1968–69	701·9	2·0

Table 4·3 uses the term 'alternative situation'. This reflects the fact that, if it were oil-fired, less theoretical capacity would be required to maintain a given generation security standard. Thus, the 35MW at the native coal and hand-won turf stations would not have been needed at all in the alternative situation, and 0·95MW oil-fired is equivalent to one MW fired by B.N.M. turf. The final column in Table 4·3 is the penultimate column multiplied by the cost per MW at the oil stations.

TABLE 4·5

Total Excess Costs

		Total excess costs (£'000)	As % of revenue
B.N.M. turf	1966–67	2206·2	8·3
	1967–68	2410·0	8·1
	1968–69	2976·2	8·6
Hand-won turf	1966–67	208·4	0·8
	1967–68	210·1	0·7
	1968–69	232·4	0·7
Native coal	1966–67	204·7	0·8
	1967–68	195·8	0·6
	1968–69	167·1	0·5
Total native fuel	1966–67	2619·2	9·8
	1967–68	2815·9	9·4
	1968–69	3375·8	9·8

Table 4·5 shows the estimate for total excess costs (it is the sum of Tables 4·2 and 4·4). From this, it can be seen that if the native fuel programme had not existed, the E.S.B. could have earned the same profit in 1968–69 with an average tariff 9·8 per cent lower than actually levied. Or, to put it another way, this programme raised average electricity prices in that year by 10·9 per cent.

It has already been pointed out that these estimates are subject to error to the extent that unused economies in generation by oil may exist and so the unit fuel and capacity costs of the hypothetical all-oil thermal system may be lower than the actual costs at the three oil stations which form the basis for these calculations. They are also defective in that other excess costs may exist. The main possibility here is transmission costs, of which there are two kinds, capital costs of the transmission system and transmission losses. Nine per cent of the units sent out are lost in the high-tension (38–220kv) system and 11 per cent in the total transmission and distribution system. In 1968–69, 61 per cent of power received at high tension was received in the Dublin, Cork and Waterford regions, where the oil-fired stations are located, whereas only 45 per cent of all units and 54 per cent of units generated thermally, came from oil stations. On the other hand, 38 per cent of all units, and 45 per cent of thermally generated units, came from native-fuel stations, whereas only 18 per cent of units received at high tension were received in the regions where the stations are located.

In the absence of locational constraints,[7] one could therefore expect a much higher proportion of the generating capacity to be located at the load centres, thus saving on the length of the heavy transmission system (and so saving in capital costs) and saving transmission losses since at a given voltage these are a function of distance. It was impracticable to measure these savings for this study, but they are undoubtedly positive, and to that extent excess costs have been underestimated.

It will be noted that the estimates vary from year to year. If the weather is wet during the turf-harvesting season, the output of Bord na Móna is restricted and so the turf stations are used less. The years 1966–67 and 1967–68 were like this, whereas 1968–69 was a good year for turf production. Thus, output from the stations served by B.N.M. rose by 53 per cent, compared with a rise of only 16 per cent in the output of all thermal stations between 1967–68 and 1968–69. Clearly, the greater the output of the native-fuel stations the greater will be excess direct costs.

Electricity consumers as such could be relieved of the burden of excess costs if the Exchequer took responsibility for the subsidy. The economic effects of this transference of the burden will be discussed later, but the point to be made immediately is that

---

[7] A constraint in the form of a requirement to use turf is a locational constraint since the value/mass ratio is so low that it is cheaper to locate stations on the bogs and transmit the product long distances than to locate at the load centres and transport the fuel.

such a transference is the only way of relieving electricity consumers of the total burden. To see this, consider the effects of two possible policy changes: the abolition of the requirement to give *preference* to turf in day-to-day operations, and then the abolition of the requirement to use turf *at all*.

First, what would be the effect if the structure of capacity remained the same but no preference were given to turf? Since hydro costs are almost all fixed, and since the Board's hydro capacity is almost totally at run-of-the-river stations, load factors at hydro stations are exogenously determined. Those at thermal stations are a matter for decision (constrained of course by outages), and the effect of preference can be seen by the following. In 1968–69 the oil stations (on average, unweighted), appeared in the load-factor ranking for thermal stations six places below where they would have appeared if merit-order principles applied. Conversely, the milled-peat stations appeared one place above and the sod-peat stations five places above where they would have been. This is not a reflection of abnormal outages since a similar pattern has occurred throughout the past six years, except in bad turf years the milled-peat stations also suffer to some extent at the expense of the sod-peat and native-coal stations.

The effect of this on costs is impossible for an outsider to determine since it would require the simulation of the Board's day-to-day operations on merit-order principles in the context of the pattern of demand and outages for a given year. However, rough estimates obtained from the E.S.B. suggest that, in the absence of preference, around £1½ million could have been saved on fuel costs. In a bad turf year such as 1967–68, the saving would be about one-third of this.

The costs of preference are not additional to the ones already estimated, since those estimates were made within the context of preference. The sums referred to in the last paragraph represent what the electricity consumer would be relieved of if preference were abandoned.

But what if the whole policy were reversed and the E.S.B. no longer had to use turf? Since, as will be seen, the least-cost policy in these circumstances is not to start afresh as if the native-fuel stations had never existed, there will be some excess costs in the new situation. The difference between these excess costs and those previously calculated is the amount by which electricity costs would fall if the native-fuel programme were abandoned. To estimate this one needs assumptions about how the Board would react to such a change in policy[8]. The calculations are made for 1968–69 only.

The hand-won turf and native-coal stations would be scrapped and the whole native-fuel system would be replaced by 250MW at turf stations converted to burn oil and 150MW of new oil-fired capacity at existing oil stations. The capital cost of conversion would be £1 million and of installing new capacity £7½ million, a total of £8½ million. The fuel cost of the new oil capacity is assumed to equal that at Great Island (the most modern station in commission at the time in question) and that of the converted turf stations is assumed to equal 1·6 times that at Great Island. The extra cost arises because of higher transport costs for oil to inland stations and because these stations would operate at low annual load factors.

It is assumed that the annual load factor at the converted stations would be 20 per cent and that at the new oil stations 50 per cent. Thus, of the output at present generated from native fuel, 40 per cent would be generated at converted stations and 60 per cent at new stations. Thus, of the 1712·9 million Kwh actually generated from native fuel, 1027·8 million Kwh would be generated at 0·406d per Kwh (unit fuel cost at Great Island) and 685·1 million Kwh would be generated at 0·65d per Kwh (i.e. 1·6 times the Great Island cost). Fuel cost would then be £3,594,200, which, when subtracted from the final figures in Table 4·1 gives excess fuel costs of £532,400.

Since the existing plant is not fully amortised, excess capital costs of £701,900 remain (see Table 4·3), and there is an additional capital charge of £850,000, the latter being the annual charge on the extra £8½ million capital. Total excess costs in the new situation would therefore be £2,084,300. The difference between this and the total shown in Table 4·5—i.e. £1,291,500—is the amount which electricity consumers would save if the native-fuel programme were abandoned. To the extent that certain costs— e.g. transmission costs—have been ignored and would be lower in the new than in the existing system, this saving is underestimated.

[8] The ensuing is based on information obtained from the E.S.B. This is not meant to be a formal statement of the Board's policy in this hypothetical situation and represents only approximate orders of magnitude, but the author is grateful for permission to use this information.

RURAL ELECTRIFICATION

Since the war there has been a special programme for connect-ing rural consumers to the system. Charges to rural consumers under this scheme are slightly higher than to urban consumers (e.g. the rural domestic tariff includes a special service charge in addition to the normal fixed charge, and for low levels of con-sumption the unit charge is higher) but the difference is insufficient to cover the extra costs of rural supply. The government pays a small grant towards the capital costs of connecting new rural consumers, but it plus the special service charge is insufficient to cover the excess, and no contribution is made at all towards the extra direct costs of rural supply.

The extra costs arise primarily because of the scattered nature of the rural load. Each unit consumed has to travel a greater distance and so unit capital distribution costs are higher, and also power has to be transmitted over larger distances at low voltage, thus increasing transmission losses.

The policy by which these extra costs cannot be reflected in the rural tariff produced annual deficits on rural account, the past three years of which are shown in Table 4·6.

TABLE 4·6

Rural Deficits

	Deficit	As % of	
	(£'000)	Urban revenue	Rural revenue
1966–67	1377·4	7·2	18·1
1967–68	1682·3	7·9	19·8
1968–69	2167·2	8·8	22·3

This table also shows that, for instance in 1968–69, this policy has produced rural prices which are lower by over 20 per cent and urban prices which are higher by almost 9 per cent than they would be in the absence of cross-subsidisation at present con-sumption levels. It is not of course implied that an increase of over 20 per cent in the average price charged to rural consumers would be either necessary or sufficient to remove the deficit. Such a large increase is bound to affect the level of consumption and whether the rural account would break even after such a rise would depend on the elasticities of demand and cost.

These estimates show the extent to which urban consumers are subsidising the rural programme. But to permit later aggregation of all the subsidies discussed here it is necessary to remove the possibility of double-counting. This exists because part of the rural deficit can be attributed to the higher costs resulting from the use of native fuel. As an approximation, one quarter of the excess costs of native-fuel utilisation can be attributed to rural account. This is shown in the first column of Table 4·7, the second column being the rural deficits less this attribution.

TABLE 4·7

'*Pure*' *Rural Deficits*

	Rural excess costs £'000	Pure rural deficits £'000	As % of total revenue
1966–67	654·8	722·6	2·7
1967–68	704·0	978·3	3·3
1968–69	844·0	1323·2	3·8

In 1968–69, the 'pure' burden of the rural deficit represented 2·9 per cent of urban revenue and 13·6 per cent of rural revenue.

RATES EXEMPTION

Under the 1927 Electricity Supply Act, the works of the new Shannon hydro scheme, which the E.S.B. was originally established to administer, were exempted from local rates. Subsequent legislation has extended this legislation to, in effect, all the Board's generating and transformer stations and transmissions lines.[9]

The number of exempt properties is enormous—the Board has over 6,000 transformer stations alone—and there could be no question of doing the relevant calculations for this study. The E.S.B. did, however, calculate that the exemptions saved them around £3½ million in 1966–67. At the time of writing, 1966–67

[9] More detail can be found in *Interdepartmental Committee on Local Finance and Taxation: Report on Exemptions from and Remissions of Rates*, Pr. 9373, Stationery Office, Dublin 1968.

was the latest year for which local taxation returns were available and these showed that the national average rates poundage was 3·7 per cent higher than in the previous year. If the same increase characterised more recent years, and if the areas in which the Board's properties are situated exhibited this increase, the minimum value of exemptions in 1968–69 was £3·75 million. It is a minimum because between 1966–67 and 1968–69 the number of exempt properties increased.

TOTAL SUBSIDIES

To summarise, the Board paid hidden subsidies in 1968–69 of £4,699,000 and received subsidies of £3,750,000. Although these cannot be regarded as offsetting each other in the economic sense (see below), the net effect upon total costs in 1968–69 was £949,000. This represented 2·8 per cent of total revenue in that year.

SOME ECONOMIC ASPECTS

The type of subsidy discussed in the foregoing sections can have two kinds of effect: upon the level and upon the distribution of welfare. This is the familiar distinction between allocation and distributional questions. The first has been the subject of an exhausting, but apparently not yet exhaustive, literature, much of which has been concerned with the question—how does one know if the level of social welfare has changed? The best known criterion produced by this literature is that if the community can move from A to B and in the process make some better off and no-one worse off, B is to be preferred to A. For a given initial distribution of income/wealth and if one wishes to pass no judgement on the merits of any distributional change produced by this process, one has, in principle anyway, a clear enough way of identifying changes in the level of welfare.

Three major problems arise in trying to apply this criterion to an actual policy situation (for instance, in judging whether hidden subsidies change total welfare). First, the Paretian criterion is so restrictive, especially if one is sceptical about Hicks/Kaldor/ Scitovsky compensation, that it is difficult to think of many policies which can be unequivocally classified as desirable or undesirable. Almost all real-life policies involve gainers and losers, and so distributional judgements can be avoided only at the cost of inaction.[10]

Second, even if one were satisfied that this were the proper criterion, it will usually be difficult to apply it in practice. Second-best theory demonstrates that, in a partial equilibrium setting, it is not enough for an improvement in welfare just to move the relevant market nearer to satisfying Paretian conditions (e.g. marginal cost pricing), although opinions vary as to what compromises can be made.[11]

Third, even if one knew what allocative and distributive effects one wanted, one may have insufficient information to pronounce upon the desirability or otherwise of a particular policy change. To take but one example, suppose there were the possibility of replacing the E.S.B./B.N.M. relationships described above with an Exchequer subsidy to B.N.M. financed by an income tax. To know whether this change would achieve the desired effect one would need to know, among other things, how this would affect the demand for electricity, how it would affect the demand for other goods and services, how it would affect the level of savings and how it would affect the supply of labour. In the Irish case anyway, there is not enough information on these points to permit cast-iron dogmatism in policy judgements.[12]

For all these reasons, no judgement is passed on the subsidies under review. This essay limits its ambition to exploring the kind of effect likely to be associated with these subsidies. The measurement of these effects is another day's work.

[10] Some writers—e.g. R. Turvey *Optimal Pricing and Investment in Electricity Supply,* London 1968, 87—prefer at least some microeconomic policy decisions to ignore distributional problems, leaving the latter to policies specially designed for the purpose. Others—e.g. B. A. Weisbrod 'Income Redistribution Effects and Benefit Cost Analysis' in *Problems in Public Expenditure Analysis.* ed. S. B. Chase, Washington D.C., 1968—believe that it is sometimes more efficient to tackle distributional and allocative problems together. Since the welfare economist thinks he has more to contribute to the latter than to the former, the views exemplified by Turvey have probably been more popular in the profession. But given that distributional problems must be faced, and given the difficulties of co-ordinating policies designed to tackle the allocative and distributional questions separately (e.g. if it is desired to leave distribution unchanged, a change in the pricing policy of a public enterprise would have to be accompanied by a change in the tax system), the Weisbrod approach may be more realistic.

[11] Again, see Turvey *op cit.* 87, for suggested compromises in the field of electricity pricing.

[12] Of course, decisions have to be made. The government cannot say 'We shall not have a tax system or impose any rules on public enterprise because we

To examine these effects it is necessary to have a point of reference with which the present situation can be compared. Two major possibilities suggest themselves: the abolition of the subsidies, and the granting of subsidies in a different way. The first is rejected since, in the native fuel and rural electrification cases anyway, abolition is not on the cards politically. Also, the abolition of the subsidy to Bord na Móna would have substantial employment effects, and the type of analysis used here proceeds on the basis of a given level of national employment. So, some other reference point is needed and the one used here is tax-financed direct grants from the Exchequer. Of course, the taxes used to finance such grants could affect the supply of resources and/or the level of employment and national income, but this possibility can be ignored on the grounds that the tax effects are highly diffused (cf. the effects of abolition, which are sectorally concentrated and which, given the structural nature of Ireland's unemployment, could influence the national employment level).[13] To keep the discussion within bounds, the direct grants are assumed to be financed from either personal income tax or turnover tax (in Ireland, the latter means a very broad-based sales tax).

For ease of comparision, the hidden subsidies can be characterised as follows. The native fuel policy is equivalent to placing a specific tax on electricity output and paying the proceeds to the producers of native fuel. The rural electrification policy is equivalent to placing a similar tax on sales to urban consumers and using the proceeds to subsidise supply to rural consumers. The rates exemption is equivalent to a surcharge on the poundage in the relevant areas, the proceeds being paid as a specific subsidy to electricity production.

The starting point is the fact that subsidies and taxes, hidden or overt, change commodity and factor prices. This distorts

do not know enough about the economy'. Only the purist in the journals can afford this luxury. The economist can properly contribute, apart from improving the information available, if he is prepared to compromise and as long as he is prepared to spell out the only vaguely supported assumptions and straight value judgements he is making (again, Turvey *loc. cit.* is a good example of this).

[13] This is admittedly a highly partial approach. Thus, if the alternative method of subsidisation were regarded as welfare improving in the cases discussed here, it is likely that they would be similarly regarded for other public sector operations. The substantial changeover to overt taxation may, but only may, have macro-economic effects.

choice (e.g. between commodities in consumption, between consumption and saving, between work and leisure, between investments of different risk and, for producers, between different production methods). It also has differential effects upon the distribution of welfare dependent on differences between individual utility functions over the relevant range.

These price changes influence the composition of consumption and production,[14] although there is insufficient information on the Irish economy to know how this composition is influenced. The net effect of the subsidies discussed is to raise the price of electricity slightly above what it would be if the income or turnover tax method were used. Since some of the additional turnover tax will fall on electricity, or if the additional income tax raises the supply price of labour, electricity prices will be higher than in the no-subsidy situation, but lower than in the present situation since the whole immediate burden would not fall on this industry. This rise in price will influence the amount of electricity consumed by an amount dependent upon the price-elasticity of demand for that product. In the short-run, this effect may be negligible because the marginal rate of substitution of other goods for electricity must be very low in the short-run. The substitution effect will be larger in the longer run, when consumers can buy equipment using other forms of energy. (In this context, it should be noted that not only would the change to the alternative method lower the price of electricity, but would raise the price of goods both complementary and competitive with electricity.)

But even in the short-run there will be an income effect: the demand for electricity can be maintained at the higher price only at the expense of reduced consumption of other goods or of reduced saving. The pattern of changes caused by the income effect will depend upon the relative income-elasticities.

Furthermore, in addition to these demand changes, there may be changes on the supply side since electricity is also an intermediate good. The effects here will depend upon the extent to which other inputs can be substituted for electricity in productive processes and the proportion of energy costs to total costs.

[14] And, sometimes, the aggregate level of consumption and output, but, as stated above, such macro effects are abstracted from here.

9

Whether all these changes would improve resource allocation is anybody's guess. If there were full employment with no structural problems and if distributional matters were ignored, or judged to be insignificant, welfare would probably be improved by the complete abolition of all such subsidies but even here second-best factors make dogmatism inadvisable. But to grant the conditional clauses in the previous sentence is to abstract from the really serious policy questions.

Although it is difficult to say how these hidden subsidies influence the level of welfare compared with the alternative situation, the pattern of distributional effects is more easy to discern—or at least the direction of these effects if not their magnitude.

The abolition of these subsidies would, as has been shown, cause electricity consumers to have to pay some $£0·9$ million less for the 1968–69 level of consumption. If the alternative is overt tax-financed subsidies, this may be an overestimate of the amount such consumers lose by the existing method since they would pay some of the extra income or turnover taxation. But if abolition were the alternative, it is an underestimate of the welfare loss of electricity consumers. Using the standard notation, if p_1 is the price in the presence of, and p_o is the price in the absence of, hidden subsidies, the foregoing estimates show differences in total outlay at a given consumption level—that is $(p_1q_1 - p_oq_1)$. But in the general case this is the minimum estimate of the welfare loss. The maximum estimate is $(p_1q_o - p_oq_o)$, and with a linear demand curve the true estimate is the arithmetic mean of the maximum and minimum. If the price-elasticity of demand were zero, $q_1 = q_o$ and so the maximum and minimum coincide and the quantities estimated in this essay would be a good measure of welfare loss. In the context of the electricity industry, this means that if demand elasticity is zero in the short-run but negative in the long-run, a substantial burden is incurred in the short-run, but some of it can be shed in the longer run by reducing consumption—e.g. by switching to other forms of energy.

Thus, unless one knows all the relevant demand elasticities one cannot measure the effects on the distribution of welfare, but at least one can say in most cases whether they are positive or negative. There is no doubt that electricity consumers suffer some welfare loss.

The hidden subsidy system also influences the distribution of income within the class of electricity consumers. The Irish income tax is progressive overall, and the turnover tax, being very broad based, is definitely regressive. The 'electricity tax' falls somewhere in between. Because of the unreliability of the data on income and expenditure by income group, no firm estimates exist in Ireland for the income-elasticity of demand for individual commodities.

Using cross-section data for 1965–66, Pratschke[15] suggests the elasticity of total consumers' expenditure with respect to personal income to be 0·75. He estimates the elasticity of expenditure on electricity with respect to total expenditure to be 0·82. If these two estimates were correct, the income-elasticity of demand for electricity would be 0·62, and one could pronounce the electricity tax to be definitely regressive. The first figure is admitted to be very doubtful but one can be sure the true figure is less than unity. If the second figure is reasonably accurate (more precisely, as long as it is less than unity), the true income-elasticity of demand for electricity is less than unity and so one can feel fairly safe in assuming this tax to be regressive.

One can therefore suggest that, taking all the hidden subsidies together, a change to direct grants financed by income tax would be egalitarian whereas a change to grants financed by turnover tax would be egalitarian if Pratschke's estimates are reasonably correct but might be inegalitarian if they were sufficiently inaccurate for the expenditure-elasticity of electricity demand to be greater than unity.[16]

So much for these subsidies as a group. They now have to be examined one at a time. The first subsidy raises the price of electricity and reduces that of turf. A change to direct grants would reduce the electricity price to the level it would be at in the absence of subsidisation and would leave the price of turf to consumers other than the E.S.B. as it is at present. The allocative and distributive implications of this have already been mentioned in the context of the subsidies as a whole.

[15] J. L. Pratschke, *Income-Expenditure Relations in Ireland, 1965–1966* Economic and Social Research Institute Paper 50, Dublin 1969.
[16] One can ignore the possibility that they are so inaccurate that this latter elasticity is so much greater than unity as to make the electricity tax more progressive than the income tax.

The rural subsidy is more interesting since its existence raises prices in one sector of the electricity market and lowers them in another. The change to grants would lower the urban price and, if paid to the E.S.B., leave the rural price as it is. In this case, however, a further possibility may be mentioned. If the argument for the subsidy is that rural consumers should have enough income to enjoy the same levels of electricity consumption as urban consumers and still have the same amount left over to purchase other goods and services or to save, this objective could be achieved by paying the grant to the consumers. From the resource allocation viewpoint this has the advantage that it would leave both urban and rural prices as they would be in the absence of any subsidy, and would extend the area of choice for rural consumers. In other words, if the argument is one of income redistribution, the objective could be achieved by an overt cash transfer direct to those it is desired to benefit. On the other hand, if the objective is to stimulate rural electricity consumption, this can be best achieved by paying the grant to the producer. Policy pronouncements have not made it at all clear which is the objective of the scheme. To the extent that there may be external economies associated with the availability of relatively cheap power in rural areas (and this could be the rationale of regarding this policy as a subsidy to agricultural mechanisation and rural industrial development) the idea of subsidising rural electricity consumption makes some sense. But if the argument is simply one of income redistribution it would seem odd to claim that the amount the beneficiaries receive should be dependent upon their consumption of electricity.

This lack of clarity about objectives also attaches to the rates exemption scheme. When this exemption was first granted, the justification given was that the ratepayers were contributing in exchange for the employment provided in their area by the construction and operation of the Shannon scheme. Such a justification could still hold with respect to generating stations, but not to transformer stations, which provide no employment, except at the construction stage. The reason for the continuation of this exemption is quite unclear. Its effect is to increase the level of rates and/or to reduce the level of consumption of public goods and services in a local area to a degree determined by the rateable valuation of E.S.B. property in that area. The allocative, and especially the distributional results are, to say the least of it, somewhat odd.[17] Using Pratschke's estimates as a guide,[18] rates are regressive with respect to income, and the degree of regression is similar to that of a comprehensive retail sales tax. Therefore, a change to grants financed from income tax would be egalitarian within a local authority area, whereas a change to turnover tax would make little difference to intra-area distribution. But also of interest is the effect on the geographical distribution of income if either change were made. Since the extra rates paid (or public services foregone) to finance the present subsidy depend on the location of E.S.B. installations, the distribution of the burden follows no pattern based on consistent concepts of equity.[19]

This is why, as pointed out earlier, it would be misguided simply to net out the subsidies and say only £0·9 million is involved. If each subsidy were similar in its allocative and distributive implications, such a procedure would be legitimate. But it is quite clear from what has gone before that, for instance, the effect of changing from the hidden native-fuel subsidy to a grant to B.N.M. is quite different from the effect of making a similar substitution for the rates exemption, even though the sums involved are almost equal.

[17] For a full discussion of the microeconomics of rates, see C. S. Shoup, *Public Finance,* London, 1969, Chapter 15.

[18] J. L. Pratschke, *op. cit.*

[19] If per capita incomes were higher in urban than rural areas, and if these installations were totally load-centre oriented, the pattern might crudely fit a principle of rich areas paying more than poor areas. But because of the locational constraints imposed by the use of hydro and native-fuel, these conditions do not exist.

CHAPTER V

The Elasticity of Demand for Irish Exports

M. J. HARRISON

INTRODUCTION

THE importance of exports to Ireland's economy need not be stressed; it is well-known. The amount of statistical analysis of Irish exports which has been published, however, is surprisingly small. The purpose of this chapter is to focus attention on the elasticity of foreign demand for Irish commodity exports and to report the results of a tentative statistical study using a number of Ireland's more important export commodities.

In international trade, many questions of practical importance hinge upon the elasticity of export demand, or of demand and supply. Economists can answer them only in a vague and indefinite manner, because they do not know the nature of the demand curve. What will be the effect of a change in transport costs or the imposition of a quota, of a new tariff or exchange rate or any other type of national action? The answers all depend in some measure on the magnitude of the elasticity of export demand of the various commodities in question.

This chapter considers the elasticity of export demand mainly in the context of the question of the effect on export earnings of an alteration in the exchange value of the Irish pound. Since the introduction of the Central Bank Bill, which amongst other things gives the Irish government the freedom to change the parity of the Irish pound independently of Great Britain, and despite an 'unequivocal' statement on the subject by the Minister for Finance, there has been some debate as to whether a unilateral devaluation of the Irish pound is or is not a likely prospect in view of the present economic difficulties.[1] This chapter establishes

[1] For a provocative assessment of some of the pros and cons of the debate see, for example, 'Devaluing the Irish pound?', *Development,* June 1970, 3–4.

a criterion of an action such as unilateral devaluation. The study of the elasticity of export demand also provides a method on the basis of which effective price policies for individual export commodities in Ireland's international trade can be framed.

A measure of the effect on export receipts of an alteration in the exchange rate is provided by a somewhat unfamiliar elasticity concept, namely, the exchange elasticity of export receipts. In order to estimate this elasticity, the more fundamental price elasticity of export demand and price elasticity of export supply need to be estimated. The usual procedure when estimating these elasticities has been to regress the aggregate of all exports on one or more broad price indices and national income. This approach is fraught with several difficulties. There is the immediate difficulty of conceiving of a regular pattern of economic behaviour, or to use Marshall's term a 'law of demand', for the varied assortment of commodities which usually constitute a country's export trade. The problem is aggravated if the composition of the assortment happens to have changed appreciably during the period available for study, as would in many instances be the case. Where such a change has taken place the measurement of aggregate exports will be severely impeded by an index number problem. Moreover where some prices have risen and others have fallen in the historical data, the resulting elasticity estimate may lie entirely outside the range of variation of the individual elasticities.[2]

Instead of estimating elasticities of export demand, supply and then receipts for exports as a whole, a more appropriate technique would seem to be to estimate these elasticities for a country's individual export commodities and then to aggregate them in some meaningful way. This procedure is reasonable in that the assumption of a 'law of demand' seems much more justifiable for individual commodities than for aggregate exports treated as a whole.[3] Several variants of this approach have been put forward, but none has received very extensive application. Such an approach is adopted in this chapter, the commodities studied being limited to four of Ireland's principle exports: cattle, beef, butter

[2] F. Machlup, 'Elasticity Pessimism in International Trade', *Economia Internazionale*, III, (1950), 132.

[3] See F. B. Horner, 'The Pre-War Demand for Wool', *Economic Record*, XXVIII (May 1952), 13.

and beer. In the five year period 1965–1969 these four constituted on average 32·9 per cent of the total annual value of Ireland's commodity exports. They accounted for 32 per cent in 1968 and 28·5 per cent in 1969. Of the four, cattle and beef, of course, are by far the most important. The model which forms the basis of the study is based on the earlier work done by Horner[4]. A fairly detailed exposition of the model is given.

THE MODEL

In any country, the price of a given export commodity may be expressed in domestic currency or, by means of the exchange rate,[5] in foreign currency units. A change in the exchange rate alters the relationship between the commodity's domestic price and its price in terms of a foreign currency. In both of the limiting cases, the change will occur all in one price level, while the other remains unaffected. In the normal case, both the domestic price and the foreign price will change somewhat[6]. Irrespective of the extent of these price changes, the first effects of an increase in the exchange rate, that is of devaluation, will usually be to encourage exports and discourage imports, while the converse will be true of a decrease in the exchange rate.

Given the exchange rate r, the export price of a given commodity in domestic currency p is simply the product of the exchange rate and the export price of the commodity in foreign currency P. Symbolically, $p = rP$. In terms of elasticities, denoted by the symbol η, it follows from the product rule for elasticities[7] that $\eta_{pr} = \eta_{Pr} + 1$, where η_{pr} is the exchange elasticity of export price in domestic currency and η_{Pr} is the exchange elasticity of export price in foreign currency. The term η_{pr} is normally positive and the term η_{Pr} is normally negative. This difference in signs means that these two elasticities are in effect complementary: the more a change in the exchange rate affects the export price in

[4] F. B. Horner, 'Elasticity of Demand for the Exports of a Single Country' *Review of Economics and Statistics,* 34 (November 1952), 326–342.

[5] The exchange rate is defined as the price of a unit of foreign currency in terms of domestic currency units. For example if one American dollar equals, say, eight shillings and one penny in Irish money, the exchange rate would be 0·4 for Ireland, expressed with respect to the American dollar.

[6] See, for example, C. P. Kindleberger, *International Economics,* 3rd ed., Illinois, 1963, 73–74.

[7] See R. G. D. Allen, *Mathematical Analysis for Economists,* London 1964, 254.

foreign currency, for example, the less it will affect that price in domestic currency.

The response of the export price of a commodity to a given alteration in the exchange rate is complicated by the fact that a currency revaluation will result in a general, albeit not uniform, movement in the prices of all export commodities, and these price changes will affect both the amount of the given commodity supplied by exporters and the amount demanded by consumers in the export market. The change in export supply, y, and the change in export demand, x, with respect to changes in the exchange rate, must be equal, therefore

$$\frac{\delta y}{\delta p}\cdot\frac{\delta p}{\delta r} + \sum_k \frac{\delta y}{\delta p_k}\cdot\frac{\delta p_k}{\delta r} = \frac{\delta x}{\delta P}\cdot\frac{\delta P}{\delta r} + \sum_k \frac{\delta x}{\delta P_k}\cdot\frac{\delta P_k}{\delta r}, \quad (1)$$

where $k\ (= 2, 3, \ldots n)$ refers to the other export commodities. The first term on the left hand side of equation (1) denotes the change in the export supply of the given commodity due to the change in its price resulting from a change in the exchange rate. Each of the other terms on the left hand side of the equation represents the change in the export supply of the given commodity due to the change in the price of some other commodity resulting from the change in the exchange rate, and it will be large only if the price change is large and the commodity is a close substitute for the given commodity. The terms on the right hand side of equation (1) denote similar effects on the export demand for the given commodity. If equation (1) is rewritten in terms of elasticities, the exchange elasticity of export price in foreign currency can be derived from it and written as follows:[8]

$$\eta_{Pr} = \frac{\eta_{yp} + \sum_k \eta_{yP_k}(1 + \eta_{P_k r}) - \sum_k \eta_{xP_k}\eta_{P_k r}}{\eta_{xP} - \eta_{yp}}, \quad (2)$$

where η_{yp} is the elasticity of export supply of the given commodity with respect to its domestic price, and is normally positive; η_{yP_k} is the cross-elasticity of export supply of the given commodity with respect to the export price (in domestic currency) of the kth export commodity, and is normally negative; $\eta_{P_k r}$ is the exchange elasticity of the export price (in foreign currency)

[8] The derivation of this expression is given in Appendix I.

of the kth export commodity, and is normally negative; η_{xP_k} is the cross-elasticity of export demand for the given commodity with respect to the export price (in foreign currency) of the kth export commodity, and is normally negative; and η_{xP} is the price elasticity of export demand of the given commodity.

An alteration of the exchange rate will also simultaneously affect the price of imports and this could have several repercussions on the export market. For example, a devaluation raises import prices in domestic currency as well as export prices, resulting in changes in the cost of imported raw materials and in the profitability of industries producing import substitutes. Both of these influences tend to modify the supply elasticity of export goods. A devaluation also causes changes in the domestic prices of imported finished goods and to the extent that a given export commodity is a substitute for such goods, the elasticity of home demand for the export commodity will be modified. If these repercussions were likely to occur with respect to any given commodity, they would have to be allowed for in the scheme of equation (2). In the case of the Irish export commodities considered in this study—cattle, beef, butter and beer— there would be an effect on the prices of the machinery which is imported for use in the production of the commodities, but the repercussions which would follow would be unlikely to have a significant influence on the state of the respective export markets and so it was not thought necessary to modify equation (2) to allow for the effect of variations of the exchange rate on import prices.

Since quantity bought is a function of price, the exchange elasticity of export demand, η_{xr}, is given by the product of the price elasticity of export demand and the exchange elasticity of export price. This is so because elasticities obey the same function of a function rule as derivatives.[9] For the reasons given above, this expression also will be complicated by additional terms containing cross-elasticities. Thus

$$\eta_{xr} = \eta_{xP}\eta_{Pr} + \sum_k \eta_{xP_k}\eta_{P_kr} \qquad (3)$$

Since revenue equals the product of price and quantity, the exchange elasticity of export receipts for a given commodity is

[9] Allen, *op. cit.*, 253.

simply the sum of the exchange elasticity of export price and the exchange elasticity of export demand. Symbolically,

$$\eta_{Rr} = \eta_{Pr} + \eta_{xr}.$$

Subsitituting (2) and (3) for the two terms on the right hand side of this relation, the exchange elasticity of export receipts can be written:

$$\eta_{Rr} = \frac{\eta_{yp} + a - b + (\eta_{xP})^2 \eta_{Pr} - \eta_{xP}\eta_{Pr}\eta_{yp} - b\eta_{xP} - b\eta_{yp}}{\eta_{xP} - \eta_{yp}}, \quad (4)$$

where a and b stand for the two cross-elasticity terms appearing in the component expressions.

However, equation (4) is not a suitable means of estimating the exchange elasticity of export receipts because the cross-elasticities which constitute a and b, though measurable in principle, are not practically measurable. The limited examination of such cross-elasticities which has been done in similar studies for other countries suggests that they are small, and too small to be measured by any feasible method. It docs not follow though, that their combined effect will be negligible. They will be more important where the export goods are close substitutes in the export market, or in the output plans of their producers, and where they supply a large proportion of the export market.[10] To obviate this difficulty equation (4) needs to be modified so as to contain only terms which are measurable. As η_{yP_k} and $\eta_{P_k r}$ will nearly always be negative and $(1 + \eta_{P_k r})$ and η_{xP_k} positive, the term denoted by a will nearly always be negative and the term denoted by b will nearly always be positive. Therefore, the term in equation (4) containing a is positive and all of the terms containing b are negative, since the denominator is negative. If these terms are omitted from the equation, what remains can be interpreted as the upper limit of η_{Rr} as long as $(b + b\eta_{xP} + b\eta_{yp})$ is not less than a. In any case, the smallness of the cross-elasticity terms means that they will not appreciably alter the size of η_{Rr} as calculated using only the measurable terms. The following equation, therefore, would seem to be the most appropriate expression to use:

$$\eta_{Rr} = \frac{\eta_{yp}}{\eta_{xP} - \eta_{yp}} + \eta_{xP}\eta_{Pr},$$

[10] Horner, *op. cit.,* 330.

or, using (2) with the a and b terms suppressed,

$$\eta_{Rr} = \frac{\eta_{yp} + \eta_{xP}\eta_{yp}}{\eta_{xP} - \eta_{yp}}. \tag{5}$$

Measurement of the exchange elasticity of export earnings for any given commodity, then, involves basically the determination of two elasticities: the price elasticity of export demand, η_{xP}, and the price elasticity of export supply, η_{yp}. Detailed consideration now needs to be given to these two elasticity concepts, and particularly to how they can be measured.

PRICE ELASTICITY OF EXPORT DEMAND

First the price elasticity of export demand. At any given level of export price p, the amount x_i of a given commodity exported from a given country to the ith country in its export market is equal to the total import demand D_i for the commodity in the ith country less the amount supplied S_i by other exporting countries. Symbolically,

$$x_i \equiv D_i - S_i.$$

Therefore,

$$\frac{\delta x_i}{\delta p} = \frac{\delta D_i}{\delta p} - \frac{\delta S_i}{\delta p}.$$

In terms of elasticities,

$$\eta_{x_i p} = \left(\frac{\delta D_i}{\delta p} - \frac{\delta S_i}{\delta p}\right)\frac{p}{x} = \frac{D_i}{x_i}\eta_{D_i p} - \frac{S_i}{x_i}\eta_{S_i p} \tag{6}$$

Thus with respect to each importing country of the export market the price elasticity of export demand for a commodity from a given country depends on the proportion of imports it supplies, $\dfrac{x_i}{D_i}$, the price elasticity of import demand, $\eta_{D_i p}$, and the price elasticity of export supply of the commodity in the competing exporting countries, $\eta_{S_i p}$. The supply elasticity term is actually made up of as many terms as there are competing exporting countries, each with its own values of S_i and $\eta_{S_i p}$.

However, the scheme of equation (6) is not realistic as it stands and requires some change before it can be used for empirical work. One important complication of the real world for which allowance must be made is the influence of the costs which account for the difference between the landed price of a commodity in an importing country and its original export price. The price elasticity of import demand in each of the importing countries of the export market is not expressed in terms of the price p, for p is the export price in the given exporting country while the price elasticity of import demand for a commodity is derived in terms of its import price, a price which includes tariffs, insurance payments and transport costs. If the import price in the ith country is denoted by $(p + t_i)$, where the term t_i stands for tariffs, insurance and transport costs per unit, then the import demand elasticity term becomes $\eta_{D_i(p + t_i)}$. However, as the import price is a function of the export price, the price elasticity of import demand can be expressed in terms of the export price by multiplying it by the elasticity of landed price with respect to export price. If the tariff is an ad valorem one and transport and other costs are negligible, the elasticity of $(p + t_i)$ with respect to p is unity and $\eta_{D_i(p + t_i)}$ remains as it is. If the tariff is specific and transport and other costs are substantial, the elasticity of $(p + t_i)$ with respect to p is less than unity, appearing as the ratio $\dfrac{p}{(p + t_i)}$, and when $\eta_{D_i(p + t_i)}$ is multiplied by this elasticity it is adjusted accordingly.

The individual supply elasticities which constitute the price elasticity of export supply in the competing exporting countries also need modification because they too are expressed, not in terms of the export price p, but in terms of their own particular export price levels for the commodity in question. If the export price level in the jth competing country is denoted by p'_j, then the corresponding price elasticity of export supply becomes $\eta_{S_{ij}p'_j}$. However, $\eta_{S_{ij}p'_j}$ can also be expressed in terms of p, if it can be assumed that there exists a functional relationship between p and p'_j, by multiplying $\eta_{S_{ij}p'_j}$ by the elasticity of p'_j with respect to p or vice versa whichever is the case depending on the existence and direction of such a functional relationship. Horner takes for granted, it seems, that p'_j simply equals $p + t_i - t'_{ij}$, where the term t'_{i_j} represents the specific tariff and transport cost per unit between the jth competing country and the ith importing country.[11] If this was in fact so, $\eta_{S_{ij}p'_j}$ could be made to refer to p instead of to the price in the competing country by multiplying it by the ratio $\dfrac{p}{p + t_i - t'_{ij}}$ which is less than unity if t_i is greater than t'_{ij} and greater than unity if t_i is less than t'_{ij}.

Unfortunately, there are objections to this approach. One criticism is that it unrealistically assumes that each importing country has only one import price for the given commodity. It would assume, for example, that in the United Kingdom the landed price of Irish beef was identical to the landed price of beef from the Argentine and beef from Australia. In fact, of course, this is not so. Another criticism is that it assumes that any change in the export price p, tariffs and other costs remaining constant, would always be exactly matched by an equivalent change in the export price level in every competing country. This, too, is unnecessarily unrealistic. A better approach is to hypothesise and estimate the relationship, if any, between the export price level p and the export price level in each competing country, and then use this estimate to compute the elasticity which is required to adjust the supply elasticity term.[12] Thus

$$\eta_{S_j p} = \eta_{S_{ij} p'_j} \frac{p}{p'_j} \frac{\delta p'_j}{\delta p}, \text{ if } p'_j = f(p)$$

and

$$\eta_{S_j p} = \eta_{S_{ij} p} \frac{p'_j}{p} \frac{\delta p}{\delta p'_j}, \text{ if } p = g(p'_j)$$

(7)

Of course, such estimation might indicate that Horner's approach would have provided a satisfactory approximation to a particular 'reaction function' and the corresponding elasticity correction factor,[13] but in general this would not be the case.[14]

[11] Horner, *op. cit.*, 329.

[12] This approach has been used by K. H. Imam, 'Export Demand Elasticities for Pakistan's Jute Trade', *Bulletin of the Oxford Institute of Economics and Statistics*, vol. 32, No 1 (February 1970), 49.

[13] This would be so, for example, where p'_j is estimated as a linear function of p with a slope very close to unity.

[14] The discussion concerning the possible interdependence between price levels in competing exporting countries introduces the question of the existence of oligopolistic price behaviour in the export trade in a given commodity, but this question was not pursued in this study.

Subject to the above discussion there emerges a slightly different equation for the export demand elasticity, $\eta_{x_i p}$, which is as follows:

$$\eta_{x_i p} = \frac{D_i}{x_i} \eta_{D_i(p+t_i)} \frac{p}{p+t_i} - \frac{S_i}{x_i} \eta_{S_i p}. \tag{8}$$

The apparently unchanged supply term on the right hand side of equation (8) is now the weighted arithmetic mean of the modified[15] supply elasticities in the competing exporting countries, the weights being the amount each competing country supplies to the *i*th importing country.

There will be an expression like (8) for each country importing the commodity from the given country. A measure of the overall price elasticity of export demand for the commodity in the export market can be derived by computing a suitably weighted average of these expressions. Thus, taking a weighted arithmetic mean,

$$\eta_{xp} = \frac{\sum_i D_i \eta_{D_i(p+t_i)} \dfrac{p}{p+t_i} - \sum_i S_i \eta_{S_i p}}{\sum_i x_i}$$

or, more simply,

$$\eta_{xp} = \frac{D}{x} \eta_{Dp} - \frac{S}{x} \eta_{Sp}, \tag{9}$$

where η_{xp} is the overall price elasticity of export demand for the given commodity from the given country, η_{Dp} is the weighted average of the modified price elasticities of import demand for the commodity in the importing countries, the weights being the respective quantities imported by the importing countries, η_{Sp} is the weighted average of the modified export supply elasticities in the competing exporting countries, and x, D and S are the total amount of the commodity exported by the given country, the total amount demanded by the export market and the total amount supplied by other exporting countries respectively.

[15] Modified in accordance with (7).

Since demand elasticities are presumably negative and supply elasticities presumably positive the two terms on the right hand side of equation (9) reinforce each other, giving a range of negative values which the price elasticity of export demand may take. The limits of this range correspond to the theoretical polar cases of monopoly and perfect competition. At one extreme, a country which has a world monopoly of a commodity simply faces the demand curve of the world market and has a price elasticity of export demand which equals the price elasticity of world demand. At the other extreme, a country which supplies a negligible proportion of the export market faces a perfectly elastic demand curve and has a price elasticity of export demand for its product of minus infinity. Between these extremes, where a country supplies more than a negligible part of the export market, it can be readily seen from equation (9) that the price elasticity of export demand for its product will be greater than the price elasticity of demand in the export market for the commodity in general the smaller the proportion of the market it supplies and the larger the response of other supplying countries to a change in its export price.

It will be obvious that several other market complications exist which should be incorporated in equation (9) if it is to be even more realistic. One of these complications concerns the definition of the export market, which is not necessarily the entire world, but in general only one particular group of countries linked by trade in the given commodity. Other groups trading in the commodity may be separated from the first and each other by various market impediments such as transport or tariff barriers. An independent country or group may of course be a potential influence on the extent of the export market: if the price goes high enough it may begin to sell to the export market; if the price goes low enough it may begin to buy from it. Another complication stems from the fact that products of different origins are by no means homogeneous in the eyes of the consumer, not even staple agricultural products. Buyers are not indifferent, for example, between Argentinian beef and Australian beef or Irish butter and Danish butter. Finally, some of the commodities exported by a country may be fairly close substitutes both in the export market and in production. Ideally the scheme of equation (9) should allow for these additional complications. However, in this study no such allowances were made. This is not to say that no attempt was made to make allowances. With regard to the failure to allow for contingent extension of the export market, the recent available data for the Irish export commodities studied suggest that the omission is justifiable. With regard to the failure to allow for the different attitudes and tastes of consumers concerning similar products from different countries, the attempt to make allowance, by incorporating into the expression for the elasticity of import demand for each commodity an estimate of the elasticity of substitution between imports and home production of the commodity, proved fruitless[16] and so was abandoned in favour of an assumption of perfect substitutability, or consumer indifference, between similar commodities from different origins. The possibility of substitution between commodities was ignored. There is little more that can be done about these omissions than to draw some reasonable conclusion about their effect on the ultimate findings, and since these complications can only reduce the price elasticity of export demand for a given country's product, estimates made using equation (9) can be interpreted as upper limits.

Equation (9) then represents the 'ideal', but such an approach could not be followed in this study owing to the lack of the relevant data for many of the countries in the export markets of the Irish goods considered, and in any case because of the sheer size of the task involved. Instead a single country approach was adopted. By this is meant that for each commodity studied an import demand elasticity was measured for only one importing country, the choice of country depending on its being a major importer and having adequate data. This single elasticity was then used as an estimate of the average import demand elasticity of all importing countries. The United Kingdom was the obvious country to choose for this. A similar expedient was used to estimate the average export supply elasticity of all the competing exporting countries, but rather than select a major export competitor, it was assumed that, for any given commodity, the export supply elasticity in every competing country was the same as the Irish export supply elasticity, η_{yp}. Consideration of how η_{yp} is estimated follows presently.

The form of the equation actually used in the study to estimate

---

[16] See below p 140.

the price elasticity of export demand for each commodity is as shown:[17]

$$\eta_{xp} = \frac{D}{x} \, \eta_{D_{UK}(p+t_{UK})} \, \frac{p}{p+t_{UK}} - \frac{S}{x} \, \eta_{yp} \left(\qquad \right), \quad (10)$$

the term in parentheses being chosen according to the conditions set out at (7) above.

PRICE ELASTICITY OF EXPORT SUPPLY

Now the price elasticity of export supply. When a change in the domestic export price level occurs, the resulting change in the amount of a given commodity supplied for export by a country is made up of the change in total output of its producers, the change in home sales due to the change in price, and the change in home sales due to the change in income caused by the change in production. The second of these component changes normally operates in the opposite direction to the other two. For example, if domestic export price increases, the resulting increase in the amount supplied equals the increase in the total output, plus the decrease in domestic sales due to the higher price, minus the increase in domestic sales due to the higher income.[18] Symbolically,

$$\frac{\delta y}{\delta p} = \frac{\delta O}{\delta p} - \frac{\delta H}{\delta p} - \frac{\delta H}{\delta N} \frac{\delta N}{\delta p}, \quad (11)$$

where y, as before, is the volume of exports of the given commodity, O is the total output of the commodity, H is the domestic consumption of the commodity, and N is the national income.

The change in national income is equal to the change in producers' receipts multiplied by a constant scale factor, the multiplier, which gives the ultimate change in national income after all ensuing rounds of spending have taken place. Since receipts is the product of price per unit and the number of units

---

[17] This equation corresponds to Imam's equation (3) in Imam, *op. cit.*, 50, and it is similar to equation (2) in Horner, *op. cit.*, 329. Horner's equation is confusing as the subscripts in the two right hand side elasticity terms relate to the wrong price level.

[18] Horner, *op. cit.*, 331.

sold, the change in national income with respect to the domestic price level of the commodity may be written:

$$\frac{\delta N}{\delta p} = \mu \frac{\delta(pO)}{\delta p} = \mu O(\eta_{Op} + 1).$$

Hence, in terms of elasticities, equation (11) becomes:

$$\eta_{yp} = \frac{O}{Y}\eta_{Op} - \frac{H}{Y}\eta_{Hp} - \frac{\mu H p O}{Y N}(\eta_{Op} + 1)\eta_{HN}, \quad (12)$$

where η_{Op} is the domestic price elasticity of total home supply, η_{Hp} is the domestic price elasticity of home demand, η_{HN} is the income elasticity of home demand, μ is the multiplier, and the other terms are as previously defined. As η_{Op}, η_{Hp} and η_{HN} are presumably positive, negative and positive respectively, from equation (12) it can be seen that the domestic price elasticity of export supply η_{yp} will be larger the more elastic is total supply and the less of it is exported, the more elastic is home demand, and the smaller are the elasticity of domestic demand with respect to national income and the producers' receipts as a proportion of national income.

The concepts from which η_{xp} and η_{yp}, and hence η_{Rr}, are derived are all, in principle at least, capable of measurement. It is now only a very short step to derive an economy's overall exchange elasticity of export receipts. There will be an expression like equation (5), which can be evaluated using (10) and (12), for each export commodity of the given country, and since it refers to revenue, aggregation is a simple matter. The exchange elasticity of total export earnings may be expressed as a weighted average of the exchange elasticities of export receipts of the individual commodities, the weights being the receipts from the export of the individual commodities. Thus

$$\eta_{(\Sigma R)r} = \frac{\delta(\sum_k R_k)}{\delta r} \frac{r}{\sum_k R_k} = \frac{\sum_k R_k \eta_{R_k r}}{\sum_k R_k}. \quad (13)$$

MEASUREMENT OF THE ELASTICITIES

Using the model presented above, it is necessary to know several things prior to estimating the elasticity of export demand and the elasticity of export supply and from these the exchange elasticity of export receipts for a commodity from a given country. First, the elasticity of import demand for the commodity must be estimated for an importing country whose elasticity of import demand is a close approximation to the weighted average of the elasticities of import demand of all importing countries. It has already been mentioned that from Ireland's point of view Great Britain was the most appropriate importing country to choose for this. In 1969, for example, 67·1 per cent of all Irish exports went to Britain.[19] In particular Britain bought 61·2 per cent of Irish cattle exports, 59·0 per cent of Irish butter exports and 59·6 per cent of Irish beer exports. Secondly, the price elasticity of total output, the price elasticity of demand and the income elasticity of demand for the commodity in the given country, Ireland, must be estimated. Thirdly, the elasticity of export price of the commodity with respect to the export price of the same commodity in a major competing country has to be estimated. Fourthly, various 'scale factors' have to be derived. For the elasticity of export demand these are the proportion of the export market for the commodity occupied by the given country, the ratio of the quantity supplied by competing exporting countries to the quantity supplied by the given country, and certain price ratios such as $\dfrac{p}{p+t}$. For the elasticity of export supply they are the three ratios involving some or all of the values of total output, domestic consumption and total exports of the commodity, the price of the commodity and the multiplier.

IMPORT DEMAND ELASTICITIES

The determination of import demand elasticities has been a subject of theoretical and empirical study for several decades. Two broad methods have been developed for estimating these elasticities, namely, direct and indirect methods. Direct methods vary from simple static models, in which imports are expressed

---

[19] That is, excluding Northern Ireland.

as a function of current values of various independent variables, to the more sophisticated dynamic models involving stock adjustments and distributed lags of varying degrees of complexity such as those associated with the names of Houthakker and Taylor,[20] Turnovsky[21] and Houthakker and Magee.[22] Indirect methods usually involve some variant of Yntema's formula[23] which requires related demand and supply functions to be estimated instead of a single import relation.

There is little evidence to say which of these two methods is better. Some writers have used direct methods and others indirect methods to estimate import demand elasticities, but only one attempt, by McAleese,[24] has so far been made to compare estimates from the two approaches. McAleese's results indicate a surprising degree of conformity between the estimated elasticities considering the different assumptions underlying the alternative methods. It would seem that convenience alone should determine the choice of method to use.

In this study the price elasticity of import demand in the United Kingdom for each of the four Irish commodities selected was determined using an indirect method based on the following expressions:[25]

$$\eta_{D(p+t)} = \frac{\eta_{Hp}\left(1 + \dfrac{1}{r}\right) - \dfrac{1}{r}\eta_{Op}\left(1 + \dfrac{m_{Hp}}{\sigma}\right)}{1 - \dfrac{(\eta_{Op} - \eta_{Hp} + m_{Op})}{r\sigma}}, \quad (14)$$

where σ is the elasticity of substitution between imports and home production of the commodity, r is the ratio of imports to

[20] H. S. Houthakker and L. D. Taylor, *Consumer Demand in the United States 1929–1970*, Boston 1966.

[21] S. J. Turnovsky, 'International Trading Relationships for a Small Country: the case of New Zealand', *Canadian Journal of Economics*, (November 1968), 772–790.

[22] H. S. Houthakker and S. P. Magee, 'Income and Price Elasticities in World Trade', *Review of Economics and Statistics*, (May 1969), 111–123.

[23] T. O. Yntema, *A Mathematical Reformulation of the General Theory of International Trade*, Chicago 1932.

[24] D. McAleese, *A Study of Demand Elasticities for Irish Imports*, Economic and Social Research Institute, Paper No. 53, Dublin 1970.

[25] M. FG. Scott, *A Study of United Kingdom Imports*, Cambridge 1963, 89.

home production and the elasticity terms are as previously defined except that here they relate to Great Britain.

A common definition of the elasticity of substitution between imports and home production of a commodity is as follows:

$$\sigma = \frac{\Delta \log M - \Delta \log O}{\Delta \log P_M - \Delta \log P_O}.$$

In words, the elasticity of substitution is simply the difference between the relative change in the quantity demand of imports, M, and home production, O, divided by the difference between the relative change in their prices. Defined in this way, σ is usually negative, the arithmetically smallest value being η_{Hp},[26] and the largest being infinity for perfect substitutes. Using this definition the elasticity of substitution between the United Kingdom's imports and home production of each of the commodities was examined with a view to using equation (14), but as was mentioned in the discussion of the model, the import elasticity term in equation (9) was not in fact modified to allow for the different tastes of consumers concerning similar products from different countries. This was because there was marked variation in the estimates of σ for all of the commodities depending on the length of the period chosen as a basis for calculating the various relative changes. For example, arithmetical values of the elasticity of substitution between Great Britain's imported and home produced butter varied between 1·26, using changes over the nine-year period 1958–1967, and 29·36 using changes during the one-year period 1967–1968. These findings were not unexpected; it is well-known that substitution elasticities are difficult to estimate correctly. It was felt that very little reliance could be placed on any of the estimates and so the idea of using equation (14) as it stands was abandoned.

It was assumed instead that imports of a commodity are a perfect substitute for domestic production of that commodity, that is, that the elasticity of substitution equals minus infinity. This was not such a bad assumption to make since the estimates of σ for each commodity were in general quite high suggesting that the substitutability between imports and home production

[26] *Ibid*, 83.

of each commodity is quite high. With regard to beef, the assumption of a very high degree of substitutability between the United Kingdom's imports and home production is in accordance with Stone's findings.[27] It is clear that as σ approaches $-\infty$, equation (14) tends towards

$$\eta_{D(p+t)} = \eta_{Hp}\left(1 + \frac{1}{r}\right) - \frac{1}{r}\eta_{Op}, \qquad (15)$$

a much simpler expression.[28] Like equation (14) this expression for the price elasticity of demand for imports is useful in that it shows the elasticity $\eta_{D(p+t)}$ is made arithmetically greater if either η_{Hp} or η_{Op} is made arithmetically greater (η_{Hp} is usually negative, remember) or if the ratio of imports to home supply, r, is reduced. It shows how a change in the price of imports exerts a double effect on the demand, by reducing consumption and by stimulating home supply. As the magnitude of σ affects $\eta_{D(p+t)}$ in direct proportion, the relation represents the upper limit of $\eta_{D(p+t)}$. Although the equation takes supply repercussions into account, there are many other repercussions it does not take into account.[29] However, it would seem to be an improvement over the methods used to estimate $\eta_{D(p+t)}$ in previous studies similar to this one which did not attempt to allow for repercussions of home supply on the elasticity of import demand, but rather approximated the latter directly using a price elasticity of demand derived from an analysis of consumption expenditure in the importing country in question.

The adoption of equation (15) as the measure of price elasticity of demand for imports meant that estimates of the price elasticity of demand and the price elasticity of supply in the United Kingdom for each commodity had to be obtained first of all. Although the demand for food and drink has been the subject of study of several econometricians, many of whose estimates

[27] R. Stone, *The Measurement of Consumers' Expenditure and Behaviour in the United Kingdom 1920–1938*, I, Cambridge 1954, 331.

[28] This particular expression is derived by Scott, *op. cit.*, 87–89; it is similar to that derived by other writers: see, for example, D. MacDougall, *The World Dollar Problem*, 1957, and M. E. Kreinin, 'Price Elasticities in International Trade', *Review of Economics and Statistics*, (November 1967), 515.

[29] See M. FG. Scott, 'Interdependence and Foreign Trade', *Oxford Economic Papers*, Volume 9 No 1, (February 1957), 88.

have been in close agreement with each other, it was decided to
estimate anew the various demand and supply elasticities using
data which was as up to date as possible. The corresponding
existing estimates, where available, were looked upon as a standby
in the event of the new estimates proving unacceptable for
whatever reason, and as a basis for comparison.

The demand equation chosen was as follows:

$$\log H = a + b \log P_d + c \log N_d + u, \tag{16}$$

where H represents domestic consumption, P_d relative price,
obtained by deflating price by the wholesale price index, and
N_d real national income, obtained by deflating income by the
consumer price index. The coefficients a, b and c represent the
estimates of theoretical parameters obtained by ordinary least
squares, and u represents the regression residuals. The constant a
depends on the choice of units. As the demand equation is of
linear logarithmic form the estimates b and c are constant elasticity estimates; they are independent of the choice of units. b is
the required estimate of the price elasticity of demand, η_{Hp}.

Since equation (16) contains no trend term, b can be interpreted
as an estimate of the long-run price elasticity of demand, and
long-run elasticities would seem most pertinent in the present
context. For a change in the exchange rate will not usually affect
the balance of payments immediately, but only after some period
sufficiently lengthy to allow for the market's response. On the
other hand, when estimating price elasticities of export demand
with a view to using them as a basis for price policy, short-run
elasticities would seem more applicable. For this reason regressions
were also carried out using a demand equation which included a
trend factor, but results consistently showed that the difference
between 'long-run' and 'short-run' estimates of the elasticities were
generally very small and not statistically significant. The estimates
obtained using the equation having fewer variables were preferred
on statistical grounds.

The estimated price elasticities of demand for cattle and beef,[30]
butter, and beer in the United Kingdom are given in Table 5·1.

[30] Because some of the data for cattle were not available, separate estimates
for cattle could not always be calculated. Where this was so the corresponding
estimates for beef were assumed to apply equally to cattle. The estimates under
the heading 'Cattle and Beef', therefore, are strictly estimates for beef only, but
the assumption made seems reasonable.

The data used were annual figures for the period 1958–1968. A complete statement of the estimates and regression statistics is given in Table A of Appendix II. Stone's estimates of these elasticities have been included in Table 5·1 for purposes of comparison.

TABLE 5·1
Price Elasticity of Demand—UK

	Cattle & Beef	Butter	Beer
η_{Hp}*	−1·23	−0·25	−0·33
	(6·09)	(3·48)	(3·65)
Stone	−0·11	−0·37	−0·87

*The number in brackets under each elasticity estimate is the corresponding *t*–value, the number of times the parameter contains its standard error.

The demand equations fit the observations quite well. The values of the parameters are plausible and in most cases significant. All of the estimated price elasticities of demand are significant at the one per cent level. However, except for butter, they differ markedly from the corresponding values obtained by Stone. Stone's estimates were made using data for the period 1920–1938; since then prevailing conditions have changed; a demand curve is not immutable. In the case of beef, as the OECD points out,[31] the rise in prices, moderate and not sufficient to discourage demand until 1963, became sharper in 1964, and demand declined appreciably in a number of countries. The OECD states that in fact a shift in demand occurred, more or less pronounced, in favour of other kinds of meat, especially pigmeat and mutton, whose prices have risen to a lesser degree than those of beef. In the United Kingdom, where mutton and beef are of comparable importance, consumers change their buying habits more easily than consumers in countries, like France, for example, where beef consumption is very high compared to the consumption of other meats. It is thought that the difference between the elasticity estimates for beef given in Table 5·1 is largely a reflection of this change. In the case of beer, a commodity with low substitutes, the difference in the estimates is thought to be due to the increase in income per head since the 1930s, the price elasticity of demand varying inversely as income per head.

[31] OECD, *The Market for Beef and Veal and its Factors*, Paris 1967, 24.

The supply of the commodities in the United Kingdom was assumed to be given by

$$\log O = a' + b' \log P' + c' \log C' + d' T + u,$$

where O represents total production or supply, P' lagged price, C' lagged factor costs and T time in years. The lagging system used for both price and factor costs was as follows:

	t	$t-1$	$t-2$	$t-3$
w_t	0	0·5	0·3	0·2

where w_t is the weight used to distribute the long-run elasticity of supply over previous years.[32] A short lagging system was chosen because it is less costly in terms of degrees of freedom. An index of gross domestic product at factor costs was used as a surrogate for the general level of factor costs, C, which was used because indexes of the particular costs incurred by the producers whose supply was in question could not be obtained. The trend factor is meant to account for a variety of influences on the level of production, such as applications of new techniques, shifts in the supply of factors of production, and the weather. It was assumed that supply changed at a constant proportional rate due to these factors during the period covered by the data. The coefficients a', b', c' and d' represent the estimates of the various theoretical constants obtained by least squares, b' being the estimate of the supply elasticity, η_{op}. The symbol u represents the regression residuals.

Because the results for butter were so poor using equation (17), an alternative supply function was hypothesised and fitted to the data for that commodity. The alternative specification was of the same general form, but unlagged price and cost variables were used instead of lagged variables. There was a considerable improvement in the results. The price elasticity of supply estimate was significant at the 1 per cent level whereas before it had not been significantly greater than zero, and the coefficient of determination, 0·15 in the first regression, improved to 0·86. These

---

[32] When the government has guaranteed minimum, or sometimes fixed prices for some period ahead, this model would seem less appropriate. However, it is unlikely that this affects the elasticity estimates to any great extent.

new results were the ones used in the subsequent calculations for butter.

The estimated price elasticities of supply for the four commodities are given in Table 5·2. Again, the data used were annual figures. Table B in Appendix II gives a detailed statement of the results of this regression analysis.

TABLE 5·2

Price Elasticity of Supply—UK

	Cattle & Beef	Butter	Beer
η_{op}	1·20	1·44*	0·26
	(1·56)	(2·47)	(1·24)

*Estimated using unlagged price and cost variables.

The fit of the cattle and beef equation is not nearly so good as the fit of the supply of butter and the supply of beer equations, but the estimated price elasticity of supply of cattle and beef, although only significant at the 10 per cent level, seems plausible. There was no noticeable improvement in results using the alternative formulation of the supply function in this case. The estimated elasticity of supply of beer also seems plausible as far as its sign is concerned, but, like cattle and beef, is only significant at the 10 per cent level. As has already been said, the estimate of the price elasticity of supply of butter is highly significant.

Using the demand and supply elasticity estimates given in Tables 5·1 and 5·2, and values of r, the ratio of imports to home production, obtained by taking arithmetic means of the ratios of quantities of imports and home production of the respective commodities for the years 1966 to 1968, the United Kingdom's price elasticity of import demand for each commodity was calculated using equation (15) and is given in Table 5·3.

TABLE 5·3

Price elasticity of Import demand—UK

	Cattle & Bee,	Butter	Beer
$\eta_{D(p+t)}$	−9·12	−0·39	−12·13
$\eta_{D(p+t)}$ using Stone's estimates of η_{Hp}	−4·36	−0·68	−23·47

Bearing in mind that these estimates are upper limits by dint of the assumption made regarding the substitutability of imports and home production of the commodities, and that they may well be substantial overestimates, they do suggest that in the United Kingdom the import demand for beef and for beer is price elastic and for butter price inelastic. This reflects the fact that the ratio of Britain's imports of butter to its home production of butter is large relative to the same ratio for beef and for beer.

PRICE ELASTICITY OF TOTAL DOMESTIC OUTPUT AND DEMAND AND INCOME ELASTICITY OF DEMAND

The price elasticity of total domestic output, the price elasticity of demand and the income elasticity of demand of the commodities in Ireland were estimated using demand and supply equations identical in form to those used to estimate demand and supply elasticities in Great Britain. Also, the data used were for the same time period. The relevant estimates are given in Tables 5·4, 5·5 and 5·6, For more detail on the regression results, reference should be made to Tables A and B of Appendix II. Included in Table 5·4, which gives the estimates of the price elasticity of demand, are similar estimates for Ireland made by Hart and Walsh, and in Table 5·5, which gives the income elasticities of demand, are additional estimates for Ireland made by Murphy, Hart, Leser, Pratschke,[33] Walsh[34] and the F.A.O.[35] Considerable work has been done on consumer expenditure in Ireland and estimates of income or expenditure elasticities are numerous.

The fit of the equations to the Irish data is in general fairly good. Only the supply equations for beef and beer have coefficients of determination below 0·8. Of the estimates of price elasticity of demand only that for butter is significantly greater

[33] J. Pratschke, *Income Expenditure Relations in Ireland 1965–66*, Economic and Social Research Institute, Paper No 50 (Dublin 1969). Pratschke gives the income elasticity estimates made by Murphy, Hart and Leser on page 13, but the figures he gives for Hart's and Leser's estimates for butter are in fact their estimates for margarine.

[34] B. Walsh, 'Economic Aspects of Alcohol Consumption in the Republic of Ireland', *Economic and Social Review,* (October 1970), 115–138.

[35] Food and Agricultural Organisation of the United Nations, *Agricultural Commodities: Projections for 1975 and 1985,* Rome 1967, Table 1·8. The FAO estimates of the income elasticity of demand for beef and butter in the United Kingdom (0·3 and 0·1 respectively) are also in very close agreement with the corresponding estimates given in Table A of Appendix II,

TABLE 5·4

Price Elasticity of Demand—Ireland

	Cattle & Beef	Butter	Beer
η_{Hp}	−0·08 (0·39)	−0·65 (2·10)	0·25 (0·87)
Hart	−1·05	0·74	—
Walsh	—	—	0·09

TABLE 5·5

Income Elasticity of Demand—Ireland

	Cattle & Beef	Butter	Beer
η_{HN}	0·50 (7·60)	0·13 (0·25)	0·42 (1·51)
Murphy	0·58	0·07	0·98
Hart	0·47	0·08	—
Leser (1964)*	0·57	0·19	0·87
Pratschke	0·79	0·13	1·79
Walsh	—	—	0·52
F.A.O.	0·5	0·0	—

*Leser also made estimates in 1962.

TABLE 5·6

Price Elasticity of Supply—Ireland

	Cattle & Beef	Butter	Beer
η_{Op}	0·97 (1·04)	1·36 (1·99)	0·15 (1·03)

than zero. The estimate of the price elasticity of demand for beer has a positive sign, but it is not significant. Walsh, using a similar demand equation, with price and income as independent variables, has reported a similar result for beer.[36] All of the

[36] Walsh, *op. cit.,* Table 8, 132.

estimates of the income elasticity of demand seem plausible and display a close correspondence to the several existing Irish estimates. The income elasticity estimate for beef is in very close agreement with the other estimates and is highly significant. The income elasticity of beer seems relatively low compared with the other figures for beer, although it is not very different from Walsh's figure; it is significant only at the 10 per cent level.

On the basis of its t-value the estimate of the income elasticity of butter is the poorest estimate, but it is in remarkable accord with the other estimates. The supply elasticities are plausible in that they are all positive, but only that for butter is significant at the 5 per cent level.

ELASTICITY OF EXPORT PRICE AND COMPETITION

In estimating the elasticity of export price with respect to the export prices in competing exporting countries, the major competing countries chosen as representative of all competing countries were for beef Argentina and for both butter and beer Denmark. The aim was to select the most important competitors in terms of total exports of the commodity, but the actual choice of countries depended on the availability of suitable data. In the case of beer, for example, Germany was the country initially selected, but an export price series was only available for all alcoholic beverages for that country and so Denmark, for which an export price series for beer was available, was substituted in place of Germany.

For each commodity a reaction function was hypothesised denoting the relation between Irish f.o.b. price and the competitor's f.o.b. price. In order to make the foreign f.o.b. prices directly comparable with the corresponding Irish f.o.b. prices, foreign currency figures were converted into pounds sterling on the basis of IMF year-end par values and suitable quantity conversion factors were used. Thus Danish f.o.b. prices for butter in Kroner per metric ton were expressed, like the Irish f.o.b. price for butter, in pounds per hundredweight. The choice of the 'price-leader' and 'price-follower' was based on the size of the respective volumes exported by the two competitors, the dominant competitor being assumed to be the 'leader'. In the price reaction function for beef, Ireland was assumed the leader; in the functions for butter and beer, Denmark, which exports more than three and a half times as much butter as Ireland and about one and a quarter times as much beer, was assumed the leader.

Two equations were specified for the reaction functions. One was a simple linear relation, the other was an equation of the constant elasticity type, linear in logarithms. Both equations were fitted to the data for each commodity using ordinary least squares, the elasticity being derived from the one which fitted the observations better. The estimated functions and the corresponding elasticities are given in Table C of Appendix II. The elasticity estimates which were chosen for use in the study are shown in Table 5·7.

TABLE 5·7

*Elasticity of Export Price with Respect to
Competitor's Export Price*

	Cattle & Beef	Butter	Beer
$\dfrac{p'}{p} \cdot \dfrac{dp}{dp'}$	0	1·03 (10·37)	0·78 (2·93)

One is immediately struck by the results for butter. The price equation fits the observations well and the reaction parameter of unity, which is the elasticity estimate in the logarithmic formulation, is highly significant. Indeed, it was apparent from the two f.o.b. price series used in the regression that corresponding Irish and Danish f.o.b. butter prices are almost identical and that there is a remarkable consistency in movement between the Irish and the Danish prices. In this case, Horner's assumption would have been quite appropriate. The equation chosen for beer is not such a good fit, but the reaction coefficient is significant at the 1 per cent level. The elasticity for beer was calculated using this coefficient and the arithmetic means of the two price series. The elasticity for beef is not significantly different from zero, and the fit of the equation to the observations is very poor, suggesting that there is no such functional relationship between the export prices of Irish beef and Argentinian beef. This does not mean that a reaction function of a different specification would not yield a higher coefficient of determination; nor does it mean that there would not be a simple relation between the Irish export price of beef and the export price of some other supplier of beef, such as one of the Commonwealth countries, Australia or New Zealand. These possibilities, however, were not pursued; rather it was assumed that other countries' export prices for beef are independent of Irish beef export prices. This seems a reasonable assumption to make.

SCALE FACTORS

The scale factors based on the export market shares of the four Irish commodities were estimated using data from the export and import matrices in the OECD publication *Trade by Commodities*. In these tables exporting countries and importing countries are arranged in columns and rows and the sources of each country's imports and destinations of each country's exports can be read off. The approximation to the export market of each commodity consisted of all the importing countries supplied by Ireland, together with all the other exporting countries which supplied them and also all other importing countries which these competitors supplied. The ratios $\frac{D}{x}$ and $\frac{S}{x}$ which were used are arithmetic means of the ratios got using value figures for 1967, 1968 and the first six months of 1969,[37] and they are given in Table 5·8. The estimated ratios may be somewhat distorted due to the fact that export values in the matrices are given in terms of f.o.b. prices and import values in terms of c.i.f. prices.

TABLE 5·8

Market Share Scale Factors

	Cattle	Beef	Butter	Beer
$\frac{D}{x}$	3·47	11·51	15·42	5·23
$\frac{S}{x}$	2·47	10·51	14·42	4·23

The scale factor $\frac{p}{p + t_{UK}}$ for each commodity is given in Table 5·9. For butter and beer it was estimated simply as the ratio of the f.o.b. price in Ireland to the c.i.f. price in Britain,

---

[37] The 1969 figures were used on the assumption that importing and exporting in all countries of the market proceeds at a constant rate throughout the year. To the extent that this is an unrealistic assumption the ratios will be biased.

but for the other two commodities this method proved unacceptable, the resulting estimates being greater than unity, implying negative transport costs! This kind of inconsistency between statistics from two different countries is not altogether uncommon. For cattle and beef, therefore, a value for t was estimated using the through transport rate from Dublin to Birkenhead, shipping agents' fees, and in the case of cattle, the compulsory insurance charge, and it was added to the Irish f.o.b. price to give the price on import to Britain.

TABLE 5·9
Ratio of Export to Import Price

	Cattle	Beef	Butter	Beer
$\dfrac{p}{p + t_{UK}}$	0·95	0·96	0·92	0·90

The scale factors used in the equation for the price elasticity of export supply of the commodities, equation (12), were calculated using average values of Irish exports, total output, domestic consumption and price level for the three-year period 1966–1968. The 1968 figures on their own give results which differ only in the second and sometimes only in the third decimal place. A value of 1·43 was used for the multiplier. This is the reciprocal of the estimate of the marginal propensity to import and is a fairly widely used figure. The estimates of these scale factors appear in Table 5·10.

TABLE 5·10
Scale Factors for the Elasticity of Export Supply

	Cattle & Beef	Butter	Beer
$\dfrac{O}{Y}$	2·054	2·420	2·050
$\dfrac{H}{Y}$	0·343	1·326	1·009
$\dfrac{\mu.H.p.O}{Y.N}$	0·044	0·080	0·124

11

Table 5·11 brings together the results given in the previous tables and gives for each commodity the price elasticity of export demand, derived by substitution in equation (10), the price elasticity of export supply, derived by substitution in equation (12), and the exchange elasticity of export receipts which was calculated by substitution of these two price elasticities in equation (5). Table 5·11 also gives the overall exchange elasticity of export receipts for the group of commodities. This last elasticity was estimated using equation (13) the weights being the averages of the export earnings of the commodities during the period 1966-1969. Using only 1969 figures as weights makes no appreciable difference to this estimate.

TABLE 5·11

Derived Elasticities of Export Demand,
Export Supply and Export Receipts

	Cattle & Beef	Butter	Beer
η_{xp}	−40·07	−63·01	−57·91
η_{yp}	1·98	3·91	0·25
η_{Rr}	1·92	3·62	0·24
$\eta_{(\Sigma R)r}$	Exchange elasticity of export receipts from Beef, Butter & Beer		1·96

RESULTS AND CONCLUDING REMARKS

Unfortunately one can not say how much confidence can be placed in these results. One can say, however, that they are likely to be substantial overestimates on three counts. Firstly, because certain cross-elasticity terms were excluded from the expression for the exchange elasticity of receipts, equation (5), which, if they could be measured and had been included, would have reduced the exchange elasticities somewhat. The important question is: by how much would they reduce the exchange elasticities? It was suggested above (p 129) that they would reduce them only slightly. On the whole the responsiveness of Ireland's export supply of any given commodity to changes in the prices of other exports would not seem to be large. Most of Ireland's exports are not close substitutes in production. Also not many of

Ireland's exports seem to be close substitutes in consumption, so that the export demand for one commodity would not be greatly affected by changes in the price of other commodities. It would seem reasonable to conclude, therefore, that the unknown cross-elasticity terms would not reduce the exchange elasticity of export earnings to any great extent.

Secondly, the results are overestimates because a number of 'market complications' affecting the price elasticity of export demand were ignored (p 134). Without further research there is no way of knowing the extent to which these factors contribute to the overestimation, but if, for example there were strong preferences on the part of consumers in the export market for Irish exports, and this might well be the case considering the large numbers of immigrant Irish in many countries, the price elasticity of export demand and hence also the exchange elasticity of export receipts, might be considerably smaller.

Thirdly, overestimation has occured because an elasticity of substitution of minus infinity was assumed for use in the equation for the elasticity of import demand, equation (15). Again there is no way of knowing how much this affected the final estimates, but it will be recalled that the elasticities of substitution of the commodities in question were thought to be fairly large, so the overestimation may not be particularly great in this case.

The results can also be questioned on statistical grounds because a number of econometric problems which probably exist to a greater or lesser degree were not considered in the analysis. For instance, where more than one relationship exists among certain variables there is the important identification problem; demand and supply functions may shift, either on their own or simultaneously, and the elasticity estimates will be biased depending on the way the functions move in relation to one another; to the extent that sizeable observational errors are present in the independent variables of the various relations the parameter estimates will be biased; the presence of multicollinearity and autocorrelation also leads to biased estimates and the Durbin-Watson statistics calculated for the regressions did indicate in some cases that there was autocorrelation in the residuals.[38] A

[38] The calculated d–statistics could only be used as a rough guide as a minimum number of 15 observations is needed before the tables of d–values can be used, However, a count of sign changes based on Geary's Tau test (*Biometrika* 1970, 57. 7, 123). seemed to confirm the conclusions made concerning autocorrelation.

transformation into first differences would have been useful in reducing this autocorrelation. However, taking first differences is also a method of trend removal and, as was mentioned above (p 142), the removal of trend means that short-run elasticities would be estimated, while longer-run elasticities were required. Hence in this study the original data have been used rather than their first difference transformations. Failure to allow for these econometric complications is likely to lead to underestimation of parameters, but the underestimation is not likely to be sufficient to more than partially cancel the overestimation due to the use of the various simplifying assumptions, and so the final estimates in Table 5·11 are still upper limits.

In view of what has just been said, only modest conclusions can be drawn. For example, it is only possible to conclude, from the last line of Table 5·11, that a unilateral devaluation (or revaluation) of 10 per cent during the period covered by the data would have led to a rise (or fall) in export earnings from beef and butter and beer combined of 19·6 per cent *at most*. The actual rise (fall) would probably have been considerably less than this, but it is not possible to say how much less. For only one commodity, beer, is it possible to say that the alteration of the exchange rate would have resulted in a less than proportionate change in foreign currency receipts from the export of that commodity. Whether the actual change in the individual export earnings of beef and butter would have been greater of less than proportionate to the change in the exchange rate depends on how much the elasticity figures given in Table 5·11 are overestimates. The true exchange elasticity of butter is more likely than the true exchange elasticity of beef to be greater than unity.

The very high export demand elasticity estimates for the commodities suggest that their true values are also quite large. This is to be expected since Ireland is a fairly small supplier of the whole export market. When export demand for a commodity is price elastic a unit decrease in price will cause a more than proportionate increase in the quantity demanded abroad. A policy implication of this is that export earnings from such commodities can be increased by reducing export prices.

The results, therefore, would seem to justify the present concern with maintaining the price competitiveness of Irish exports, at least in respect of the commodities examined here. It should be noted that if a policy of export price reductions was effected, by export bonuses, subsidies or reductions in costs of production, for example, it would have corresponding repercussions on the price elasticity of export demand. For as Ireland's share of the export market increased, the scale factors $\dfrac{D}{x}$ and $\dfrac{S}{x}$ would decrease, and this decrease might lead to such a decrease in the export demand elasticity, that the policy can no longer be pursued with advantage. Furthermore $\dfrac{S}{x}$ might decrease enough to cause competitors to react with a retaliatory price cut. This would tend to stabilise the export demand elasticity, however. Finally, importing countries would probably react and alter their tariff policies vis-à-vis the export price policy for the goods in question pursued by Ireland.

To reiterate the point made in the introduction, the statistical section of this study is only a tentative one, meant primarily to illustrate the use of the model; it makes no pretence of econometric sophistication. Nevertheless, it seems reasonable to accept the crude estimates of the various elasticities as first approximations and upper limits, at least until more refined data and improved techniques have been used to yield more accurate results.

If these results are accepted the next step would be to apply the model to a sample of some of Ireland's more important manufactured commodity exports to provide estimates of the demand elasticities of those manufactures and, by pooling the results with the figures given above, a better estimate of the overall exchange elasticity of Irish export receipts.

APPENDIX I

DERIVATION OF EXCHANGE ELASTICITY OF EXPORT PRICE IN FOREIGN CURRENCY

Equation (1) is

$$\frac{\delta y}{\delta p}\frac{\delta p}{\delta r} + \sum_k \frac{\delta y}{\delta p_k}\frac{\delta p_k}{\delta r} = \frac{\delta x}{\delta P}\frac{\delta P}{\delta r} + \sum_k \frac{\delta x}{\delta P_k}\frac{\delta P_k}{\delta r}$$

In terms of elasticities

$$\eta_{yp}\eta_{pr} + \sum_k \eta_{yp_k}\eta_{P_k r} = \eta_{xP}\eta_{Pr} + \sum_k \eta_{xP_k}\eta_{P_k r},$$

and since

$$\eta_{pr} = \eta_{Pr} + 1,$$

$$\eta_{yp}(\eta_{Pr} + 1) + \sum_k \eta_{yp_k}(\eta_{P_k r} + 1) = \eta_{xP}\eta_{Pr} + \sum_k \eta_{xP_k}\eta_{P_k r}$$

$$\therefore \eta_{yp}\eta_{Pr} + \eta_{yp} + \sum_k \eta_{yp_k}(\eta_{P_k r} + 1) = \eta_{xP}\eta_{Pr} + \sum_k \eta_{xP_k}\eta_{P_k r}$$

$$\therefore \eta_{yp} + \sum_k \eta_{yp_k}(\eta_{P_k r} + 1) = \eta_{Pr}(\eta_{xP} - \eta_{yp}) + \sum_k \eta_{xP_k}\eta_{P_k r}$$

$$\therefore \eta_{Pr}(\eta_{xP} - \eta_{yp}) = \eta_{yp} + \sum_k \eta_{yp_k}(\eta_{P_k r} + 1) - \sum_k \eta_{xP_k}\eta_{P_k r}$$

$$\therefore \eta_{Pr} = \frac{\eta_{yp} + \sum_k \eta_{yp_k}(1 + \eta_{P_k r}) - \sum_k \eta_{xP_k}\eta_{P_k r}}{\eta_{xP} - \eta_{yp}},$$

which is equation (2).

APPENDIX II

TABLE A *Regression Results for Demand*

$$\log H = a + b \log P_d + c \log N_d$$

	United Kingdom			Ireland		
	Cattle & Beef	Butter	Beer	Cattle & Beef	Butter	Beer
a (constant)	10·24	6·01	−4·00	2·49	3·77	1·22
b (price elasticity of demand)	−1·23 (6·09)	−0·25 (3·48)	−0·33 (3·65)	−0·08 (0·39)	−0·65 (2·10)	0·25 (0·87)
c (income elasticity of demand)	0·39 (3·62)	0·16 (2·30)	0·83 (20·24)	0·50 (7·60)	0·13 (0·25)	0·42 (1·51)
R^2 (coefficient of determination)	0·85	0·78	0·98	0·87	0·91	0·93
F—value	19·81	13·85	243·65	30·01	25·54	61·35
d (Durbin-Watson Statistic)	2·44	1·64	2·19	2·08	1·00	1·05

TABLE B *Regression Results for Supply*

$$\log O \ a' + b' \log P' + c' \log C' + d'T$$

	United Kingdom			Ireland		
	Cattle & Beef	Butter★	Beer	Cattle & Beef	Butter	Beer
a' (constant)	23·67	−6·77	1·72	10·20	−8·16	5·18
b' (price elasticity of supply)	1·20 (1·56)	1·44 (2·47)	0·26 (1·24)	0·97 (1·04)	1·36 (1·99)	0·15 (1·03)
c' (cost elasticity of supply)	−5·48 (1·21)	0·40 (0·86)	−0·66 (1·83)	−1·57 (1·99)	0·83 (1·77)	−0·02 (0·04)
d' (logarithm of constant rate of change of supply per annum)	0·12 (0·95)	0·004 (0·82)	—	0·04 (0·79)	—	—
R^2 (coefficient of determination)	0·48	0·86	0·98	0·47	0·92	0·45
F—value	0·94	16·64	62·28	1·46	34·07	3·74
d (Durbin-Watson statistic)	1·71	1·67	2·52	2·56	1·97	1·44

★The estimates for butter were made using unlagged price and cost variables.

TABLE C *Regression Results for Reaction Functions*

	Beef		Butter		Beer	
	Equation I	Equation II	Equation I	Equation II	Equation I	Equation II
Estimated reaction equation	$p' =$ $-8 \cdot 72 + 1 \cdot 24\, p$ $(1 \cdot 28)$	$\log p' =$ $-1 \cdot 42 + 2 \cdot 05 \log p$ $(1 \cdot 35)$	$p =$ $-0 \cdot 86 + 1 \cdot 02\, p'$ $(11 \cdot 03)$	$\log p =$ $-0 \cdot 06 + 1 \cdot 03 \log p'$ $(10 \cdot 37)$	$p =$ $1 \cdot 32 + 0 \cdot 40\, p'$ $(2 \cdot 93)$	$\log p =$ $-0 \cdot 07 + 0 \cdot 72 \log p'$ $(2 \cdot 84)$
R^2 (coefficient of determination)	0·19	0·21	0·92	0·93	0·52	0·50
F—value	1·62	1·82	100·66	107·48	8·61	8·05
Elasticity adopted $\dfrac{p}{p'}\dfrac{dp'}{dp}$	—	0	—	1·03	0·78★	—

★ Estimated using the means of the variables.

CHAPTER VI

The Distribution of Personal Wealth in Ireland[1]

PATRICK M. LYONS

INTRODUCTION

THERE are estimates of the distribution of personal wealth for the U.S. from 1922,[2] and for the U.K. from 1911.[3] These estimates are important because they tell us about one of the most fundamental factors in life—the ownership of property. Those who own property have greater freedom to vary their consumption of goods and services than those who must rely on current income, they have greater security, more power and considerable influence on the pattern of saving and investment which is the mainspring of economic growth.[4] Moreover, such estimates of wealth ownership show property to be most unequally distributed and therefore the advantages of property accrue to only a few in society. Yet despite the importance of the distribution of wealth, there are no comparable estimates for Ireland.

In this chapter, a well tried and accepted technique is used to estimate the total personal wealth in Ireland. The distribution of this wealth is given by age and sex. In addition, the distribution of the wealth according to form of assets is calculated. The approach is not new, but some innovations have been made in the methodology which may be of interest and the information as far as Ireland is concerned is entirely original.

[1] The author would like to express his gratitude to the following for their assistance in the assembling of the data for this analysis: the Minister for Finance, Mr George Colley, the Chairman of the Revenue Commissioners, Mr James Duignan, members of the Office of the Accountant-General of Revenue, particularly Mr Sean O'Meadhra, and the staff of the Estate Duty Branch, especially Mr John Daly and Mr Patrick Ahern.

[2] R. J. Lampman, 'Changes in the Share of Wealth Held by Top Wealth-Holders, 1922–1956', *Review of Economics and Statistics,* 41 (November 1959), 379–392.

[3] G. W. Daniels and H. Campion, *The Distribution of National Capital,* Manchester 1936.

[4] A.A. Tait, *The Taxation of Personal Wealth,* Urbana, Illinois 1967, Chapter 1.

THE NATURE OF THE CALCULATION

In simple terms, deceased individuals whose estates come up for examination for estate duty purposes in a given period of time are regarded as being a representative sample of all individuals in the country. Accordingly, the wealth which forms their estates is regarded as being a representative sample of the wealth possessed by the surviving individuals in the country. Each deceased is classified by sex and age group, and the numbers of persons in each sex and age group in the country is known from the Census of Population. The total number of deceased in each sex-age group cell is expressed as a proportion of the numbers similarly classified in the whole population, The reciprocal of this proportion is then applied to the total wealth possessed by the deceased individuals, and this grossed-up amount is taken to represent the total wealth possessed by all persons in that cell in the surviving population. Let us assume that in a particular sex and age group cell that 1,000 persons died in a particular year. If there were 10,000 persons in the entire population in that cell during the year, this implies that the mortality rate was 10 per cent. If those deceased possessed total wealth between them of £1 million, it follows, according to this formula, that among the whole surviving population in that category total wealth amounted to £10 million.

This approach was developed initially by researchers, and has latterly been adopted by official agencies in many countries. Important pioneering work was performed for Britain in 1954 by Lydall and Tipping,[5] but similar estimates are now made annually by the Commissioners of H.M. Inland Revenue.[6] For Ireland, a version of the basic technique was used by Nevin[7] to estimate personal wealth in the period 1953–55. A modified version was required in that exercise because the basic classification of estates by sex and age-group was not available.

Before proceeding further, it is as well to outline some of the difficulties and limitations connected with this procedure. In all countries, the following points must be borne in mind:

    [5] H. F. Lydall and D, G. Tipping, 'The Distribution of Personal Wealth in Britain', *Bulletin of the Oxford University Institute of Statistics,* Volume 23, No 1, (1961), 83–104.
    [6] For example, see *Inland Revenue Statistics, 1970,* H.M.S.O., 1970, 176–184.
    [7] E. T. Nevin, *The Ownership of Personal Property in Ireland,* Economic Research Institute, Paper No. 1, (Dublin 1961).

(a) Generally speaking, only estates over a certain minimum value are liable to payment of duty, and are therefore subjected to close scrutiny. In addition, most individuals possess only a very small amount of wealth, and therefore detailed information is available each year for only a small proportion of total estates, and for only a very small proportion of all personal property then in existence. (In Ireland, in recent years, only an estate whose net capital value exceeded £5,000 was subject to estate duty. In 1966, according to published and derived statistics, these liable estates totalled only 1,830, forming a proportion of only 14·94 per cent of all the 12,250 estates examined that year, 5·55 per cent of the 32,956 deaths among persons aged 20 and over, and only 0·11 per cent of the total adult population of 1,724,250).

(b) There is always the possibility of evasion of estate duty due to non-declaration, or under-declaration of the value, of assets. Non-declaration may be expected with cash and moveable property, including (particularly in Ireland), livestock. Under-declaration might occur, for example, in the valuation of shares of private companies.

(c) Where relief of estate duty is allowed for gifts *inter vivos*, difficulties arise whose magnitude is hard to estimate. If gifts are made prior to a certain date before death, they are not included in the estate of the deceased at all. If the period between the gift and death is less than a certain number of years, such gifts are included in part in the estate, the proportion of the gift increasing the nearer are the events of settling and death. Nevin[8] argues that the properties concerned are subject to a dual mortality risk during this period, and may thus be over-estimated. Outside this period, however, there is only a single mortality risk, and since it is normal that gifts are made from one individual to a younger person, that mortality risk is lower than would have been the case if the gift had not been made. Some estates are therefore subject to estate duty twice in a short

[8] Nevin, *op. cit.*, 4.

period of time, whereas other estates always escape the estate duty net.

(d) Life insurance policies are included in an estate at their value after death. This may be considerably greater than their value immediately prior to death. Again, Nevin[9] considers this to provide an element of over-estimation. Lydall and Tipping[10] found, however, that there was considerable under-statement of sums assured, part of which they attributed to the fact that many life insurance policies are associated with mortgages on dwelling-houses, and the proceeds of the matured policies are used to extinguish the mortgage debt. The value of the house would be included in the estate, but not the value of the insurance policy.

(e) Finally, there are difficulties connected with undervaluation of assets due to the generosity or kind-heartedness of those responsible for the assessment. Independent assessments of the value of many assets are accepted by the officers in charge of the collection of estate duty. The alleged rapaciousness of the tax-collector does not materialise in the person of the supervisor of death duties.

In Ireland, two further difficulties are important when estate duty statistics are employed for the purposes of wealth estimation. In the first place, it is assumed that persons dying in a particular year are representative of the surviving population; if there is an undue delay in the presentation of estates for assessment, then the estates in any one year must relate to a period of time prior to that year. In a period of rising prices this leads to undervaluation of the assets in the estates. A very long time-lag obtains with regard to Irish estates, and this is commented upon further in the Appendix. Secondly, certain assets are legally entitled to be given a value below their true value for estate duty purposes. This applies particularly to assets in the form of agricultural land and Stock Exchange securities. This problem will be discussed later in this chapter.

Bearing these defects in mind, the estate duty statistics still remain the most convenient method of arriving at some estimat-

[9] *Ibid*, 5.
[10] Lydall and Tipping, *op. cit.*, 103.

ion of the total wealth in Ireland. A more comprehensive account of, and justification for, the methods employed will be found in the Appendix. For the moment, a shorter account will do.

As was mentioned earlier, previous work in this field was limited as there were no estate duty statistics classified by age and sex. Nevin,[11] in conclusion stressed that his work was hampered by the inadequacy of published data in this respect. Since that time, some estate duty data classified by sex and age-groups have been published, and Nevin must be given the credit for stimulating this activity.

Nowadays, the Revenue Commissioners publish an analysis of property by age and sex of deceased.[12] These statistics cover all estates liable for estate duty, being those whose net capital value exceeds £5,000. There is no analysis published giving, for these estates, an analysis of the net total size of the estate classified by age and sex; it is merely the components of each estate which are thus classified. No information at all is published concerning estates below the minimum level for payment of estate duty, except the total capital value of such estates presented to the Estate Duty Branch.[13]

More comprehensive information relating to both large estates (defined as those liable for payment of estate duty, i.e. above a net capital value of £5,000) and small estates had to be obtained. This was done by means of a detailed and, in some ways, comprehensive study of the basic data used in the preparation of the statistics in the Reports of the Revenue Commissioners.

The year 1966 was chosen as the base for this study, firstly because it coincided with a Census of Population, and secondly because it was assumed that this date was sufficiently far back in time to ensure that all pertinent estates would have been dealt with fully. The validity of this latter assumption is questioned in the Appendix. Since such a small number of estates above the lower limit appear each year, it was considered that it would be preferable to combine the estate duty statistics for the years 1965–66 and 1966–67 in order to include persons deceased in the two years 1965 and 1966.

---

[11] Nevin, *op. cit.*, 18.
[12] For example, *Forty-Fourth Annual Report of the Revenue Commissioners, Year ended 31st March 1967*, Pr. 9680, Stationary Office, (Dublin 1967), 116–131.
[13] *Ibid.*, 110–111.

TABLE 6·1(a) All Estates Classified by Range and Age-group—Males 1965–66 and 1966–67

Age Group

Range of Estate	20–24 years	25–34 years	35–44 years	45–54 years	55–64 years	65–74 years	75–84 years	85 years and over	Total
Under £100, Not exceeding	67	—	—	—	—	267	133	—	467
Exceeding £100, Not exceeding £1,000	67	67	267	601	1,536	2,137	2,271	1,336	8,282
£1,000, £2,000	—	67	133	67	467	534	601	133	2,002
£2,000, £5,000	—	—	67	267	868	1,002	1,002	267	3,473
Total small estates	134	134	467	935	2,871	3,940	4,007	1,736	14,224
Exceeding £5,000, Not exceeding £6,000	—	7	6	29	78	110	130	36	396
£6,000, £7,000	—	3	12	19	64	96	90	39	323
£7,000, £8,000	—	2	3	19	60	85	69	21	259
£8,000, £10,000	—	4	15	28	84	170	135	41	477
£10,000, £12,500	1	2	7	19	57	130	122	44	381
£12,500, £15,000	—	—	6	15	50	72	80	22	246
£15,000, £17,500	—	2	5	16	44	57	52	12	188
£17,500, £20,000	—	2	6	4	22	40	47	15	136
£20,000, £25,000	1*	—	6	13	30	58	60	16	183
£25,000, £30,000	—	1	2	8	29	30	50	10	130
£30,000, £35,000	—	—	3	5	9	35	28	10	91
£35,000, £40,000	—	—	1	1	14	21	7	9	53
£40,000, £45,000	—	—	—	—	6	14	14	3	37
£45,000, £50,000	—	—	1	3	3	12	17	11	47
£50,000, £60,000	—	—	—	5	10	12	14	6	47
£60,000, £75,000	—	—	1	1	5	12	12	5	36
£75,000, £100,000	—	—	1	1	8	12	10	5	37
£100,000, £150,000	—	—	—	—	3	6	14	5	29
£150,000, £200,000	—	—	—	1	—	2	7	1	10
£200,000, £250,000	—	—	—	1	—	—	1	1	3
£250,000, £400,000	—	—	—	—	—	—	—	1	2
£400,000	—	—	—	—	2	—	—	—	2
Total large estates	2	23	75	189	578	974	959	313	3,113
Total all estates	136	157	542	1,124	3,449	4,914	4,966	2,049	17,337

Age Group

Range of Estate		20–24 years	25–34 years	35–44 years	45–54 years	55–64 years	65–74 years	75–84 years	85 years and over	Total
Under £100	Not exceeding	—	—	—	67	67	67	200	134	535
Exceeding £100	£1,000	—	—	—	200	467	1,269	1,469	668	4,073
£1,000	£2,000	—	—	—	67	200	334	534	134	1,269
£2,000	£5,000	—	—	—	—	67	534	601	133	1,335
Total small estates		—	—	—	334	801	2,204	2,804	1,069	7,212
£5,000	£6,000	—	2	4	11	23	63	85	49	237
£6,000	£7,000	—	—	1	7	20	51	67	33	179
£7,000	£8,000	—	—	—	3	17	30	57	37	144
£8,000	£10,000	—	1	—	3	23	65	85	47	224
£10,000	£12,500	—	1	1	7	27	50	78	32	196
£12,500	£15,000	—	1	1	1	4	27	49	28	111
£15,000	£17,500	—	—	—	—	6	28	48	21	103
£17,500	£20,000	—	—	—	1	10	10	22	19	62
£20,000	£25,000	1*	—	2	3	8	25	28	24	91
£25,000	£30,000	—	—	—	1	7	13	26	14	61
£30,000	£35,000	—	—	—	—	1	11	23	14	49
£35,000	£40,000	—	—	—	—	3	8	11	7	29
£40,000	£45,000	—	—	—	2	4	1	2	8	17
£45,000	£50,000	—	1	1	2	—	4	8	3	19
£50,000	£60,000	—	—	1	—	1	5	9	12	28
£60,000	£75,000	—	—	—	1	1	4	12	8	26
£75,000	£100,000	—	—	1	—	2	1	6	6	16
£100,000	£150,000	—	—	—	—	2	5	4	5	16
£150,000	£200,000	—	—	—	—	—	—	3	—	3
£200,000	£250,000	—	—	—	—	—	—	1	—	1
£250,000	£400,000	—	—	—	—	2	—	1	—	3
£400,000		—	—	1*	—	—	—	—	—	1
Total large estates		1	6	13	42	161	401	625	367	1,616
Total all estates		1	6	13	376	962	2,605	3,429	1,436	8,828

*These items are excluded from subsequent calculations.

The smaller estates presented serious problems, and while an examination of the data used in the published statistics yielded useful information, it was found necessary to undertake a small sample survey which involved an examination of the actual files for those persons whose estates were scrutinised by the Estate Duty Branch, but who paid no duty upon those estates.

These two investigations produced an analysis of all estates during the period April 1965 to March 1967, classifying the number of estates in each range of value of estate by sex and age-group. This analysis is not part of the result of the inquiry into wealth, but since the results are of importance, and since they have not previously been published, they are reproduced here as Tables 6·1 (a) and 6·1 (b). The assumption in all subsequent calculations is that the combined results for both years apply to the Census of Population date in April 1966. The Tables here give the totals for both years in order to avoid the half-persons which would have resulted from dividing each item by 2 to represent a single year. The percentage distribution would, of course, remain unchanged. Finally, it must be stated that the investigation is concerned with the ownership of capital amongst those aged 20 years and over and with those whose domicile was in Ireland. All those leaving estates of any value below that age are excluded completely from this inquiry, as are all persons domiciled outside Ireland who left property in the State.

Two interesting points emerge from these Tables. In the first place, it appears that for both small and large estates, two males own capital for every one female. Secondly, the conclusion might reasonably be drawn that females tend to own capital at a somewhat later age than males. This is apparent from the larger estates, where the modal female holding occurs in the 75–84 year age group, as opposed to the 65–74 year age group for males. Even in total estates, the distribution is such as to suggest that females still hold capital at older ages than males. While the mode occurs in both sexes in the same 75–84 age group, the distribution of numbers in the adjacent age groups indicates a later modal age for females. This is a reflection of the fact that females have a longer life expectancy than males, and that there-fore there are more women alive at advanced ages (a) to own capital, and (b) to have inherited it from their deceased spouses. The concentration of capital owned by women in the older age groups results in a slightly less equitable distribution of capital between the sexes than the estate duty analysis would indicate, as is demonstrated below.

THE RESULTS

THE DISTRIBUTION OF PERSONAL CAPITAL

In addition to earlier and subsequent assumptions and explanations about the capital holdings of persons, it is assumed that all individuals who died during the years of inquiry who did not have estates falling within the scope of the inquiry held a net wealth of zero. An impression was gained, however, that if anyone left net wealth below £50, his benefitors would not present his estate to the Estate Duty Branch. If all these were added to the following estimates at £50 each they would have the effect of raising the estimates by about £56 million, or 2·5 per cent.

The total estimated values of capital owned by persons in Ireland in 1966 are shown in Table 6·2. Three individuals shown by means of asterisks in Tables 6·1(a) and 6·1(b) have been omitted in these calculations since their large estates were owned at a very early age, and their inclusion would have a seriously damaging effect upon the calculations. The inclusion of one of these resulted in the addition of over £100 million worth of wealth to one of the younger female age groups. This is the great disadvantage of the small sample size in Ireland, compared with the much larger number of estates which are analysed each year in the U.K. and other countries. To increase the number of estates would require the combination of estates over more than two years, and this would introduce other serious problems, particularly in respect of the valuation of assets. Table 6·2 shows, for each size category of net capital, the estimated number of persons in that category, together with the estimated amount of capital (to the nearest £1 million) owned by these persons. The percentage distribution of persons and capital is shown in Table 6·3, where larger net capital ranges are used.

These estimates suggest that total personal net capital in Ireland in 1966 was £2,121 million. Of this £1,518 million was owned by about 92,000 persons who possessed £5,000 or more each. Thus, while nearly 65 per cent of the population possessed no

12

TABLE 6·2

Estimated Distribution of Personal Net Capital in Ireland, 1966

		Number of Persons aged 20 or over	*Amount of net capital (£ million)*
Net Capital			
Nil		1,120,278	—
Under £100		47,120	1
Exceeding	*Not Exceeding*		
£100	£1,000	286,078	181
£1,000	£2,000	98,444	154
£2,000	£5,000	80,301	267
Total small estates		1,632,221	603
Exceeding	*Not Exceeding*		
£5,000	£6,000	15,289	84
£6,000	£7,000	10,574	69
£7,000	£8,000	7,122	53
£8,000	£10,000	13,980	126
£10,000	£12,500	10,491	118
£12,500	£15,000	6,723	92
£15,000	£17,500	5,672	92
£17,500	£20,000	4,146	78
£20,000	£25,000	5,230	118
£25,000	£30,000	2,881	79
£30,000	£35,000	2,351	76
£35,000	£40,000	1,157	43
£40,000	£45,000	745	32
£45,000	£50,000	1,813	86
£50,000	£60,000	1,218	67
£60,000	£75,000	821	55
£75,000	£100,000	1,005	88
£100,000	£150,000	457	57
£150,000	£200,000	67	12
£200,000	£250,000	80	18
£250,000	£400,000	157	51
£400,000 and over		50	24
Total large estates		92,029	1,518
Total all estates		1,724,250	2,121

capital, about 30 per cent of the population in the next category, which included amounts up to but not exceeding £5,000, possessed over 28 per cent of total capital, the remaining 5 per cent, approximately, of the population possessed amounts of £5,000 and over, and this group accounted for over 70 per cent of the total wealth in Ireland. The top 1 per cent of the population owned over 30 per cent of the net wealth. Indeed, the 0·047 per cent at the top of the distribution owned nearly 7·6 per cent of the total wealth. According to these estimates, males owned 72 per cent of the total net capital, while females possessed 28 per cent of it.

TABLE 6·3

Percentage Distribution of Adult Population and Wealth in Ireland, 1966

Net Capital		Percentage of Persons aged 20 or over	Percentage of Net Capital
Nil		64·972	0·000
Not exceeding £5,000		29·690	28·438
Exceeding	*Not Exceeding*		
£5,000	£10,000	2·724	15·650
£10,000	£20,000	1·568	17·926
£20,000	£50,000	0·822	20·477
£50,000	£100,000	0·177	9·912
Exceeding £100,000		0·047	7·597
Total		*100·000	100·000

*In this and subsequent Tables, the individual items may not add to the totals shown because of rounding.

These estimates produce values which are substantially higher than those of Nevin,[14] where he estimates total net wealth on average in the years 1953–55 to have been £792·4 million. In the intervening years there was a reasonably rapid rise in the level of prices, and undoubtedly a considerable amount of real accumulation of property. In similar ways, discussed below, there is likely to have been underestimation in both Nevin's exercise and the current one, although there is also reason to suspect a degree of overestimation in the latter analysis. Taking personal net wealth as a percentage of National Income, it stood at 251 per cent in 1966, but, according to Nevin,[15] at 178 per cent in 1953–55, as compared with 282 per cent in 1937–39, and 257 per cent in the period 1923–25. In comparison, in Great Britain in 1966, personal net wealth was 255 per cent of National Income.[16] Taking the very large differences between the economies of Ireland and Great Britain, the similarity between the ratios cannot be regarded as being in the nature of a verification of the

[14] Nevin, *op. cit.*, 8.
[15] Nevin, *op. cit.*, 8.
[16] Derived from *Inland Revenue Statistics, 1970,* 181, and *Annual Abstract of Statistics, No. 100, 1969,* H.M.S.O., 1970, 271.

present estimates of net capital in Ireland. Nevertheless, it does give grounds for arguing that they are of the correct order of magnitude.

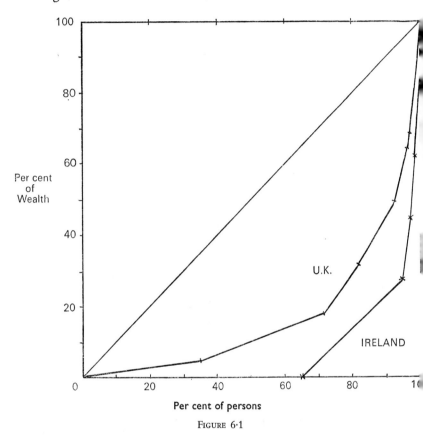

FIGURE 6·1

It is possible to draw a Lorenz curve of the distribution of wealth in order to demonstrate the inequality of that distribution. This is done in Figure 6·1. The 45° line would occur if all wealth were equally distributed. The further is the deviation of the curve from the diagonal, the greater is the degree of inequality. It is normal to compare the inequality in the distribution of wealth with the inequality of income distribution. Such an exercise is not possible in Ireland, due to the lack of information concerning income disrtibution. Some purpose is served by comparing the Irish experience with the distribution of wealth in Great Britain in 1966.[17] As may be seen from the diagram, while the distribution of wealth in Britain shows considerable inequity, the distribution in Ireland is even more inequitable. In the United States, the distribution of wealth is less unequal than that in Britain, and this curve would accordingly be inside the British one, closer to the diagonal. The unusual path of the Irish Lorenz curve along the horizontal axis is due to the assumption that no wealth at all is owned by the bottom 65 per cent of the population.

THE COMPONENTS OF PERSONAL CAPITAL

The estates of those above the estate duty exemption limit in Ireland are published in component form, classified by age group and sex, but not by size of estate, as was explained above. A similar grossing-up exercise as before was applied to each of the components for the years 1965–66 and 1966–67, except those components of the three estates excluded previously. The estimated components of personalty are shown in Table 6·4, and those of realty in Table 6·5. These are combined in Table 6·6, which shows the components of total personal capital.

TABLE 6·4

Estimated Components of Personalty in Ireland Held by Persons With Over £5,000 Total Net Capital, 1966

	Total Asset Values £million	Percentage of Total Net Personalty
1. Government and Municipal Securities—Irish	102	12·70
2. Government and Municipal Securities—Foreign	19	2·36
3. Corporate Securities—Irish	181	22·44
4. Corporate Securities—Foreign	148	18·41
5. Mortgages, Money on Bills, etc.	3	0·43
6. Household Goods	22	2·76
7. Insurance Policies	117	14·49
8. Cash in House and at Bank	139	17·30
9. Trade Assets	74	9·17
10. Other Assets	108	13·36
Total Gross Personalty	914	113·43
Reductions	108	13·43
Total Net Personalty	806	100·00

[17] U.K. source is *Inland Revenue Statistics,* 1970, 181.

TABLE 6·5

*Estimated Components of Realty in Ireland, Held by Persons With
Over £5,000 Total Net Capital, 1966*

	Total Asset Value £million	Percentage of Total Net Realty
1. Land	173	49·36
2. Houses and Business Premises	127	36·26
3. Other	58	16·45
Total Gross Realty	357	102·07
Reductions	7	2·07
Total Net Realty	350	100·00

TABLE 6·6

*Estimated Components of Total Personal Capital in Ireland
Held by Persons With Over £5,000 Total Net Capital*

	Total Asset Values £million	Percentage of Total Net Capital
1. Government and Municipal Securities—Irish	102	8·85
2. Government and Municipal Securities—Foreign	19	1·65
3. Corporate Securities—Irish	181	15·64
4. Corporate Securities—Foreign	148	12·83
5. Mortgages, Money on Bills, etc.	3	0·30
6. Household Goods	22	1·92
7. Insurance Policies	117	10·10
8. Cash in House and at Bank	139	12·06
9. Trade Assets	74	6·39
10. Other Assets (Personalty)	108	9·31
11. Land	173	14·96
12. Houses and Business Premises	127	10·99
13. Other Assets (Realty)	58	4·98
Total Gross Capital	1,271	109·99
Deductions	115	9·99
Total Net Capital	1,156	100·00
Of which:		
Personalty	806	69·70
Realty	350	30·30

It is immediately apparent that the estimate for total personal capital using the published data is much less than the total obtained by the analysis of actual estates, the totals being £1,156 million in the former case, and £1,518 million in the latter. The reasons for this discrepancy are explained in the Appendix. Of more importance in the present context is the distribution of the components of personal property, although the estimated values of the components can be used to demonstrate undervaluation of the assets themselves.

It appears from these calculations that real property in the form of land, houses, business premises, and other assets, such as farming stock and non-farm stock, and plant and machinery, accounted for 30·3 per cent of total personal investment. Of the remaining 69·7 per cent, approximately 35 per cent consisted of Irish stocks and shares, almost 21 per cent was invested in foreign stocks and shares, while 17·3 per cent was held in the form of cash in the home and at the bank, and nearly 14·5 per cent consisted of insurance policies.

In spite of the fact that no analysis is possible of the composition of estates below the exemption limit there is undoubtedly serious under-valuation in several of the components. Some of these occur because there is allowance made for the under-valuation of certain assets for estate duty purposes. There is a convention whereby agricultural property may be valued below its market value in certain circumstances. If the effect of the artificial valuation is to bring the total value of the estate below £1,000, then agricultural property (including land and buildings) may be valued in the following manner: the annual value (rateable valuation) is multiplied by 25, and from this is subtracted the redemption value of any outstanding annuity on it. The vast majority of Irish farms fall under this provision, which has the effect of often removing farms worth tens of thousands of pounds from imposition of estate duty. This analysis discloses the total value of agricultural land in the larger estates at £173 million. Even if this figure is doubled on the assumption that land in the estates below the exemption limit is valued at a similar amount, (and would account for nearly 30 per cent of all capital in those estates, which is not entirely unreasonable) this implies an average value per acre for all 11·82 million acres of arable land of only £29·3. This must seriously undervalue agricultural land, and a similar underestimation would occur in the case of agricultural buildings.

With securities in most companies quoted on the Irish Stock Exchange another artificial value is allowed, provided that the deceased was domiciled in the State. One third is deducted from the market value of these securities on the date of death, provided that the securities have been held since the date of issue or for 3 years, whichever is the shorter. In addition to Irish public corporate securities, the estimated figure of £181 million also includes shares in private companies. These latter are not allowed an artificial value, but it is reasonable to assume that assets owned by private companies are undervalued in order to arrive at a share value for estate duty purposes. At the end of 1965, the paid up capital of all Irish companies was over £275 million, while the nominal share capital was over £498 million.[18] Some of the share capital, particularly of the quoted companies, was undoubtedly owned by financial institutions such as insurance companies, but it is certain that the paid up capital underestimates the value of company paid up share capital, and it is likely that the nominal share capital does also. The underestimation is probably less than is the case with agricultural property, but it nevertheless appears to be substantial.

Other components appear to be substantially undervalued also. Indirect evidence can illuminate some of these discrepancies. Total holdings by persons of stocks and shares abroad are estimated at £167 million. In 1966, however, the C.S.O. estimates that the income from investments abroad and extern profits was £44·9 million.[19] Not all of this income was income from investment, nor did it all accrue to persons. But it appears either that the Irish are very wise foreign investors, or, more likely, that there is underestimation in the figure for total personal external investment. The estimate for insurance policies produces a figure of £117 million. Apart from considerations about life insurance policies outlined above, one further feature is of importance. This is that the insurance policies included in the estate duty returns were taken out some time, and often a con-

[18] *Statistical Abstract of Ireland, 1968,* Prl. 189, Stationery Office, (Dublin 1968) 345.
[19] *National Income and Expenditure, 1967,* Prl. 515, Stationery Office, (Dublin 1969), 74.

siderable time, before the year in question. It is certain that sums assured in the past were much smaller than those assured currently. In 1966 alone, total new sums assured in respect of life policies by both Irish companies and by non-Irish companies with business within Ireland was over £108 million.[20]

Cash in the house and at the bank is estimated at £139 million in this analysis. In March 1966, currency outstanding and current and deposit accounts (adjusted) with the Associated Banks totalled nearly £589 million.[21] A large proportion of this total was owned by companies, the government and by the banks themselves. But once again underestimation of personal holdings may be suspected, particularly when it is remembered that deposits with other financial institutions, including the Post Office Savings Bank, are not included in currency and bank deposits. Building societies alone were liable for shares and deposits of over £33 million at the end of 1966, while deposits with the Post Office Savings Bank (including the Trustee Savings Banks) and Savings Certificates totalled over £177 million at the same time.[22] Finally, houses and business premises were estimated to have a value of £127 million, but total capital formation in the form of building and construction amounted to nearly £114 million during 1966.[23] Only a small proportion of this investment comprised houses and business premises, but, given the life expectancy of these assets, the estimate from the estate duty statistics appears on the low side.

THE DISTRIBUTION OF WEALTH BETWEEN AGE GROUPS

Studies of wealth distribution in other countries demonstrate that a dis-proportionate share of total wealth is owned by the relatively small number of elderly persons, so that average wealth increases with age. Lydall and Tipping,[24] for example, estimated that average net capital per person in the 20 to 24 year age group in 1954 was £330, rising with each age group, to a maximum point of £2,310 in the 75 years and over group.

A similar investigation was undertaken with the Irish data,

[20] *Statistical Abstract of Ireland, 1968,* 345.
[21] *Report of the Central Bank of Ireland, 1967–68,* Dublin 1968, 87 and 104.
[22] *Ibid.,* 111, 98 and 99.
[23] *National Income and Expenditure,* 1966, 53.
[24] Lydall and Tipping *op. cit.,* 96.

and the results are contained in Tables 6·7 and 6·8. Table 6·7 shows the estimated distribution of all capital, while Table 6·8 shows the estimated distribution of capital where the net capital exceeded £5,000. The results demonstrate that those above the age of 55 years possess, on average, more wealth than those below that age.

But the results are surprising in that they differ from experience elsewhere. Average net capital per person in both analyses increases with age until it reaches its maximum level in the 55 to 64 year age group. It drops quite sharply in the following group, rises again in the 75 to 84 year group, and finally falls again in the final group comprising those 85 years and over.

Part of the explanation for this phenomen might lie in the fact that Ireland has a fairly large proportion of elderly people, many of whom own very little capital. Another explanation would suggest that a relatively large proportion of the elderly are in the agricultural population, since they would tend to live longer than town-dwellers, and their farms are given an artifically low valuation. In addition, this unusual distribution might suggest that the transfer of property before death was more common in Ireland than had previously been thought. One stimulus to this might be the desire of the elderly to divest wealth in order to qualify for the old age pension. Nevertheless, these results are interesting.

CONCLUSION

This paper has been a combination of an attempt to arrive at estimates of personal wealth-holding in Ireland, and a trial of the usefulness of applying techniques employed elsewhere to Irish data. The following conclusions emerge from the analysis:

(a) Estate duty statistics would suggest that total personal wealth in Ireland in 1966 was of the order of £2,121 million, which represents an average of about £1,231 per head of the adult population, and about £735 per head of the total population. It represented about 2½ times that year's National Income, and just under 30 per cent was owned by females, who comprised just less than half the total population, and more than half in age groups above 75.

TABLE 6·7 *Estimated Distribution of All Capital Between Age Groups, Ireland 1966*

Age Group	Per Cent in Group Owning Capital	Per Cent of Total Net Capital Owned by Age Group	Per Cent of Total Population Aged 20 and over in Age Group	Average Net Capital per person £
20–24 years	35·39	1·69	10·75	193
25–34 years	23·56	11·45	17·16	821
35–44 years	31·17	17·89	18·42	1,195
45–54 years	34·07	18·53	19·22	1,186
55–64 years	45·93	24·36	15·71	1,908
65–74 years	42·55	16·35	11·85	1,698
75–84 years	29·13	8·24	5·71	1,777
85 years and over	32·80	1·49	1·18	1,556
Total	35·03	100·00	100·00	1,231

TABLE 6.8 *Estimated Distribution of Capital for Those Owning More Than £5,000 Net Capital Between Age Groups, Ireland 1966*

Age Group	Per Cent in Group Owning More Than £5,000 Net Capital	Per Cent of Total Net Capital Owned by Group	Per Cent of Total Population Aged 20 and over in Age Group	Average Net Capital per person £
20–24 years	0·26	0·44	10·75	36
25–34 years	4·49	10·14	17·16	520
35–44 years	5·13	18·21	18·42	870
45–54 years	5·12	19·16	19·22	878
55–64years	7·69	24·82	15·71	1,391
65–74 years	7·60	16·59	11·85	1,233
75–84 years	7·39	8·86	5·71	1,368
85 years and over	6·58	1·77	1·18	1,325
Total	5·34	100·00	100·00	881

(b) There is substantial inequality in the distribution of wealth. Nearly two-thirds of the adult population owned no wealth, while the 5 per cent of the population at the top of the distribution owned over 70 per cent of total personal capital. This inequality is more marked than in Great Britain and the United States.

(c) Of total personal capital, just over 30 per cent is held in the form of real property. Half of this is accounted for by land, and a further third consists of houses and business premises. 69·7 per cent is held in the form of personalty, of which over one third is comprised of Irish stocks and shares, 20 per cent is held abroad in stocks and shares, over 17 per cent is held in cash and bank deposits, and insurance policies account for over 14 per cent further.

(d) The most significant conclusion of this analysis might well be the final result which showed that average wealth does not, in Ireland, rise with age, but reaches a peak in the 55 to 64 year old age group which, in turn, suggests that many gifts *inter vivos* are made.

(e) It is apparent that some assets are seriously undervalued in the assessment of estates for estate duty purposes. This affects the validity of the technique used, but alternative estimating procedures might be used to supplement the basic procedure in respect of certain items, particularly land.

(f) In the application of this technique, two sets of data were used, since the existing estate duty statistics are inadequate for this purpose. These data yielded substantially different results for the value of total net personal capital. This factor, and the basic task of obtaining data suitable for processing, suggests that improvements can be made both in the collection of estate duty and in the publication of statistics in this field.

APPENDIX

ESTIMATING METHODS EMPLOYED

THE TREATMENT OF LARGE ESTATES

Large estates, i.e. those which exceeded a net capital value of £5,000 and were therefore subject to estate duty, were subjected to two entirely different procedures. As was explained above, the existing published data do not show estates classified by size of estate, by sex and by age group. Thus Table 6·2 had to be derived by examining the basic estate duty statistics. This was done in three stages. In the first place, as each estate pays duty, either on the whole estate or upon a part of it, records are kept of such estates month by month, in the Office of the Accountant-General of Revenue. These records include the name of the deceased, the file number of the estate, the size category of the estate, the sex and age (where known) of the deceased, and the various component parts of the estate upon which duty is at that time paid. Monthly record-books are kept of the components, each record-book including only estates of a particular size category.

The aggregate records at the end of each year, which are published, form a book-keeping record of all estates which paid duty in that year. The statistics count an estate only on one occasion, that being when the first payment of duty is made. The aggregate amount of estates is, however, the total amount of estate components which become liable for duty during each year. As an estate first presents itself for payment of duty, it is classified to a certain size-range of total net estate. As further amounts are added to the estate, or as further deductions are made from it, the estate may move from one size category to another. Since estates are also classified by age group, estates may also move from the category 'Age not disclosed' to a specific age group, when the age of the deceased is revealed, after the first portion of the estate is presented for assessment of duty. These transfers of estates and parts of estates are made, in the book-keeping statistics, not only within a year, but also between years, since in many cases the full value of an estate is only discovered after the lapse of a considerable period of time, often many years. When the annual statistics are published, in some categories of net estate-sex-age group there are negative amounts, conveying the impression that, for example, males included in the net estate range £35–40,000 aged 55 to 64 in 1966–67, between them left a net capital of minus £24,820.[25] Duty was, however, paid on estates of males in that age-net capital category, although this would not be apparent from the published statistics, whereas a refund of duty would appear more appropriate.

It was therefore necessary to abstract from the monthly records details of all individuals who paid duty on any portion of their net estates during the period April 1965 to March 1967. Since an individual could appear several times during this period, and move from one net capital category to another, all the records were checked against each other from the end of that period to the begining, in order to ensure that no individual estate appeared more than once, and that each estate was as near the end of its career in paying estate duty as possible, so that it would be in the correct size category during that period.

The estates thus abstracted included those which appeared only in this two-year period, those which had appeared first in some previous period and subsequently appeared between April 1965 and March, 1967, and others which first appeared in this period, but which also continued to appear in later years. It is appropriate at this stage to outline the manner of distribution of these files which were included in the analysis as regards the year in which the estate first appeared to pay duty.

TABLE 6·9

Date of Opening of Estate Duty Files, 1965–66—1966–67

Date of Opening of File	Number of Files
Before 1959	214
1960–1963	475
1964	522
1965	1,643
1966	1,519
1967	349
After 1967	7
Total	4,729

[25] *Forty-Fourth Annual Report of the Revenue Commissioners,* 129.

Certain essential points must be explained about these files. In the first place, the date given is that on which the file was first opened in the Estate Duty Branch. This is in almost all circumstances sometime after death, but the generally accepted time-lag of three months between death and presentation of the estate was found to be wildly inaccurate. The time-lag in the case of Irish estates is considerable. In many cases it amounts to years, in some cases to decades. In addition, there is also a considerable time-lag between the assessment of duty and the payment of duty, when it appears in the statistics of the Accountant-General for Revenue. This is due to the very low penalty rate of interest (4 per cent) on unpaid duty. The explanation for the opening of files post-1967 in the Table arises from the fact that a file which originally appears to be liable for nil or a low duty, is later given a new file number if it appears liable for much larger duty. This may be done in a later year, and subsequent operations in this investigation required that note should be taken of the final file number. Finally, a few of the very early files were opened before a person's death to pay duty in advance of death on a trust settlement in order to commute succession duty, and this file number normally remained even if that person died many years later. But the important feature of this investigation was that the estates often referred to a period before the period of inquiry, and that their components were accordingly undervalued.

The details thus collected gave no details of the age of the deceased in about 10 per cent of the cases, and in others left the domicile open to doubt. Accordingly the records of the Estate Duty Branch were examined to ascertain domicile from their card-index, and to discover age from the actual files themselves. Finally, where age was still unknown, an investigation was undertaken into the Register of Deaths at the Office of the Registrar-General at the Custom House. A very small number of estates were excluded because ages were not ascertainable, some dying before 1921 or outside the country, or because domicile could not be established due to the difficulty of locating their files.

For each estate in the period of inquiry, there were available details of the deceased as regards sex, age, size of net estate and domicile. Those domiciled outside Ireland were excluded as were those aged under 20. This produced the distributions of Tables 6·1(a) and 6·1(b). The estates in each age, sex and size group category were then grossed to give up to the number of persons in each category for the total population using the method described above, excluding three specific estates. Elsewhere, it has been the practice to assume that those owning wealth had a better mortality experience than the general adult population and to use, as did Lydall and Tipping,[26] mortality rates applicable to the upper social classes in the population. Since there are no such social class mortality rates available in Ireland, the mortality rates used were those of the general adult population. This may result in some under-estimation. The mid-point of each net capital class was taken as the mean value of all estates in that class, apart from the last, open-ended class, where the actual values were taken. These calculations produced the distributions of Tables 6·2, 6·3, 6·7 and 6·8.

The distribution of the components of personal wealth in Ireland, which appear in Tables 6·4, 6·5 and 6·6, were obtained by applying the same mortality rates to the data published by the Revenue Commissioners, again with the exception of three estates. The data for the years 1965–66 and 1966–67 were combined, as before, and this resulted in the disappearance of all but one of the negative classes described above. The data include estates of persons domiciled outside Ireland, and these are also included in the grossing-up procedure, but the final results exclude those included in the data as 'Age not Stated'. Each component was grossed up by the relevant mortality rate.

These two approaches produced total large estates of 4,729 and 3,752, and estimates of total personal net wealth of £1,518 million and £1,156 million respectively. These differences arise from the fact that in the first case all estates appearing in the estate duty returns were included at their full value in the analysis. In the latter instance, estates were only included as to the actual amount of the estate which appeared in the statistics for both years. Neither approach is entirely justified. Ideally, with sufficient resources, all estates of persons who died in a particular year should be examined individually. This would avoid the pitfalls of giving possibly too much weight to every estate which appears in the published statistics, as is done in the main analysis, and of giving too little weight to all estates which began appearing

---

[26] Lydall and Tipping, *op. cit.*, 98–100.

in the period of inquiry, as is done by using the statistics published by components.

If it is assumed that the estimation of the value of small estates is reasonable, then the percentage of National Income which arises from each of the analyses is 251 in the main inquiry, as opposed to 208. The examination of underestimation in the latter, component, estimation, must produce a conclusion that that estimation of total net personal capital is well below the true value, and possibly that the other estimate is below also. The main justification for the basic approach used here is that the estate duty statistics are seriously out of date, although, it must be added, this provides no justification for the arbitrary approach of including all items in the statistics at their full value. Apart from the very large difference in the value of total estates, the two approaches show reasonable similarities in other respects, as is demonstrated in Table 6·10.

TABLE 6·10

	By main analysis Percent	By component analysis Percent
Percent of personal capital owned by:		
(a) Males	72·0	70·2
Females	28·0	29·8
(b) Persons aged		
20–24 years	0·44	0·89
25–34 years	10·14	9·53
35–44 years	18·21	18·23
45–54 years	19·16	21·99
55–64 years	24·82	22·59
65–74 years	16·59	16·56
75–84 years	8·86	8·53
85 years and over	1·77	1·68

THE TREATMENT OF SMALL ESTATES

No details are published concerning small estates, except the total net capital value of the estates upon which no duty was paid. Until September, 1965, however, daily records were kept of the number of estates in this category classified by amount of net capital. The records show the number of estates in each category, together with the total capital value of those estates. These can be used to demonstrate the number of estates in each category, and the mean value of those estates. The actual statistics

13

of estates in the year 1964–65 were added to twice the values revealed in the period April 1965 to September 1965 which were assumed to correspond to the year 1965–66, in order to produce the figures given in Table 6·11.

TABLE 6·11

The Distribution of Small Estates, 1964–65—1965–66

Range of Estates		Number of Estates	Percentage of Estates	Total Value of Estates £	Average Value of Estates £
Under £100		2,017	9·3	129,015	64·0
Exceeding	Not Exceeding				
£100	£1,000	11,644	54·0	5,286,354	454·0
£1,000	£2,000	3,055	14·2	4,370,025	1,430·5
£2,000	£5,000	4,860	22·5	15,328,211	3,154·0
Total		21,576	100·0	25,113,605	1,164·0

These figures suffered from two disadvantages from the point of view of the present exercise. They gave no details as regards age and sex of the deceased, nor could they be considered up-to-date. Accordingly, a small sample survey was undertaken basically to determine the age and sex structure of the estates below the exemption limit. A sample of 150 estates from each of the years of the inquiry was chosen by means of a table of random numbers. Eventually over 800 estates were examined because the original estates in the sample did not give the age of the deceased, or, more generally, were estates of a person who died in a period before that of the inquiry. 321 estates were used in the final analysis, 213 of them left by males, and the remainder by females. It was assumed that the total number of estates in this category was, for the period 1965–66 to 1966–67, four times the number of estates in the six months between April and September, 1965, since no figures were available for the number of small estates in the relevant period. The sample estimates grossed up in this fashion are reproduced in Tables 6·12 (a) and 6·12 (b).

These estimates appeared to be in reasonable accord with those presented in Table 6·11. That the percentage in the lowest category is smaller, and the average capital values in each class are greater, is a reflection of the fact that each of the estates which form the basis for Table 6·12(a) and (b) apply to those who died in the period of the inquiry, and that the estates in Table 6·11 are those of persons who died in an earlier period.

Each of the figures in the basic distribution was then grossed up using the mortality rates of the general population, to produce some of the statistics in Tables 6·1(a) and (b), and these were then multiplied by the average net value of estate figures in order to complete Tables 6·2 and 6·3. Those persons whose estates were not examined were assumed to have possessed net capital of zero.

TABLE 6·12(a)

Estimated Distribution of Small Estates, 1965–66—1966–67, Males

Range of Estates		*Number of Estates*	*Percentage of Estates*	*Average Net Value of Estates* £
Under £100		469	3·3	55·3
Exceeding	Not Exceeding			
£100	£1,000	8,278	58·2	473·0
£1,000	£2,000	2,006	14·1	1,487·6
£2,000	£5,000	3,471	24·4	3,298·8
Total		14,224	100·0	1,291·8

TABLE 6·12(b)

Estimate Distribution of Small Estates, 1965–66—1966–67, Females

Range of Estates		*Number of Estates*	*Percentage of Estates*	*Average Net Value of Estates* £
Under £100		534	7·4	53·0
Exceeding	Not Exceeding			
£100	£1,000	4,075	56·5	457·8
£1,000	£2,000	1,269	17·6	1,593·4
£2,000	£5,000	1,334	18·5	3,366·5
Total		7,212	100·0	1,165·8

CHAPTER VII

The Brain Drain Question

M. O'DONOGHUE

THE 'brain drain'—the international migration of highly-qualified personnel such as university graduates—is a subject which has attracted considerable interest in many countries in recent years. Given that the general level of emigration is high by international standards it might be expected that this topic would also have had extensive discussion in Ireland. Such is not the case. The only substantive treatment and discussion appears to have been that associated with the publication of Professor Lynn's paper in 1968.[1]

The present treatment deals primarily with the theoretical aspects of this brain drain question. One reason for this emphasis is that the available data are sparse and inadequate to support any extensive empirical treatment. A second reason is that the Lynn paper and subsequent discussion focussed primarily on empirical issues so that the present exposition may serve as a useful complement to this earlier debate.

The sequence adopted here is to explore first the general economic consequences of brain drain movements—this is done in the first section. In the second section the characteristics most appropriate to Irish circumstances are selected on the basis of the earlier discussion. In the final section there is a brief discussion of the Lynn paper in the context of this present analysis.

GENERAL ECONOMIC ASPECTS

THE COMPREHENSIVE MARKET CASE

The first setting in which to consider the brain drain is that which might be expected to prevail in conditions where the market system was functioning in a fully comprehensive manner. This term is intended to describe a situation where all economically relevant transactions are satisfactorily dealt with by a market

---

[1] R. Lynn, *The Irish Brain Drain*, Economic and Social Research Institute, (Dublin 1968).

process of allocation, or put differently it is assumed that all of the complications which arise from instances of market imperfections or failure are absent. Of the more important among these latter factors it is specifically assumed that full employment prevails and that no externalities (unpriced inputs or outputs) are present. The implications posed by such factors as these are considered later.

The first point to clarify is whether any migration would arise in this comprehensive market case. The assumption is that the market is functioning in an equilibrating manner, which implies that all price and quantity movements, including migration, represent part of an adjustment process designed to achieve some new equilibrium. Once an equilibrium is attained there is no reason to expect that migration or any other adjustment, would continue. Historically, it is not difficult to envisage how migration would arise. Initial differences in national endowments of natural resources or in population structures could result in isolated countries having different income levels. The introduction of transport for example would then facilitate movements of both goods and people as part of an adjustment process to eliminate the initial income disparity. There is no reason to expect that migration movements forming part of this or any similar adjustment to a given imbalance, should take any systematic pattern, hence there is no reason to expect that some areas would be continuing sources of emigration, or that others would be regular recipients of immigrants.

Some other sources must be sought then to explain continuing migration in such a system. Two relevant possibilities appear to exist. The first of these is differences in birth rates between countries. The effects of such a difference can be illustrated by assuming two economies similar in all respects save that one has a higher birth rate than the other. This would mean that complete equilibrium would not be achieved because each time equality in income per head were approached, and migration ceased, there would then be a tendency for income per head to start declining in the area with the faster population growth, and this tendency would in turn stimulate emigration in order to restore equilibrium between the two areas. However, while this factor of differential birth rates could persist for some time, it must be questioned whether it would be a permanent feature. The emigration would mean that there would be a growing fraction of the population in the immigrant area who had originated in the high birth rate country. This would suggest that the initial differences in birth rates should eventually disappear, unless of course, it is assumed that the differences are in some way related to geographical location and not to other influences.

The second source of continuing migration could be differences between countries in their rates of technical progress. If one again assumes two relatively similar countries, this time differing in their technical progress, it would be expected that the one with the faster rate of progress would tend to have the faster income growth. This tendency towards greater income levels would attract migrants from the slow-growing area. If these trends were continued for a very long period one might logically expect that the faster growing area would develop a tendency towards relative over-population which could offset its technical leadership and eliminate any gap in income levels. For a continuing tendency towards income disparity it might be more realistic to combine an assumption of faster technical progress with one of a lower birth rate in order to preserve conditions facilitating immigration to the specified area.

These problems need not however be pursued. It may be assumed that there are conditions in which migration could occur in a comprehensive market system. The next question to explore is whether, and to what extent, a brain drain would form a part of such migration. Grubel and Scott[2] in a discussion of this question suggest several reasons for expecting a greater incidence of brain drain than of other categories of migrants. Since highly-qualified personnel earn above average incomes, the transport and other costs of movement between countries will represent a lesser obstacle to them than they do to other groups. Secondly, they are less likely to find language differences an obstacle, given their above average educational and ability levels. They are also likely to be better informed about the opportuinities and environment in different countries, and are less likely to have insular attitudes.

One may accept this reasoning and now examine what economic effects if any would result. First, since the movement

[2] H. Grubel and A. Scott 'The Immigration of Scientists and Engineers to the U.S. 1949–61', *Journal of Political Economy* (August 1966).

is voluntary one may reasonably conclude that the individuals concerned gain as a result of their migration. The more complex issue is the nature of the effects on the rest of the economy. Given the assumption of a fully-functioning market there is no reason to expect any such effects on the production side of the market. The departure of an individual will mean the loss of his output, but also a saving of equal amount in payments made to him for this output. It is on the consumption side that the possibility of a gain or loss arises. There is no reason to expect that lifetime consumption will equal lifetime production for every individual: on the contrary, numerous imbalances would be probable. If such imbalances were randomly distributed throughout the entire population then there would be no need to explore them in a brain drain context. Such however is unlikely to be the case. Highly qualified personnel earning above average incomes would be expected to on balance accumulate wealth through savings. Savings release resources to the rest of the economy when they are made. If they have not been fully realised before death—that is, the individual dies possessing some net wealth—then lifetime consumption is less than the value of his lifetime production, and there is a temporary net gain by the recipients of the savings, and a permanent net gain by the inheritors of the wealth.

A second way in which there may be a transfer of benefits to other people would be by way of unilateral unrequited transfers. The need for such transfers would exist because in a comprehensive market setting there would be many people with no or very inadequate incomes, so that voluntary community measures to alleviate poverty would be expected, and these would call for transfers of both money and time on the part of more fortunate groups. It seems reasonable to bracket highly qualified personnel among the more fortunate groups.

If it is accepted that on balance brain drain categories would be donors of gifts and wealth to the rest of the community, the next issue is whether migration would lead to any geographical redistribution of these transfers. At the one extreme it is unreasonable to expect a complete redistribution in the sense that the location of transfers is uniquely tied to the location of the donors. Gifts in the form of emigrants' remittances have been a significant by-product in the case of emigration from Ireland for example. But equally at the other extreme it would be unwise to expect a zero redistribution of transfers in the sense that the country of origin continues to be the recipient of all gifts.

The more probable expectation is that some transfers would be made to the emigrant's country of origin, but also that some, especially those associated with participation in the voluntary programmes for poverty alleviation, would take place in the country of residence. If this is so then there is some net loss to the country of origin as a result of emigration by highly-qualified personnel, since in the absence of such migration all of these transfers would have been retained in the country of origin.

THE EXTERNALITIES CASE

The second case to explore is that which extends the analysis to take account of the presence of externalities. Such unpriced inputs and outputs can arise in a number of ways, but for present purposes they can be thought of in terms of those cases where the market fails to function in a comprehensive manner, or where it has been replaced by government financing or provision. In the brain drain context, one of the most frequent examples of an externality concerns the education of the migrants involved. It is usually the case that governments meet a large portion of the costs of education. Hence the college graduate, for example, receives a benefit considerably greater than the cost he himself has borne. If he then emigrates, he leaves 'owing a debt to society'. Education is not the only such externality which may arise. The young emigrant may also have received medical, welfare or other governmentally-provided benefits, for which less than full payment was made. Externalities may also arise in the private sector. It is frequently contended, for example, that wage rates do not reflect the actual value of the work done by each person. If some are overpaid, the presumption is that others are underpaid. The usual examples suggest that on balance it is the highly qualified who are underpaid; for instance the doctor whose skill saves many lives, or shortens illnesses, but who receives less than the value of the savings he effects, or the research worker whose discoveries provide benefits far beyond his salary receipts.

It is also possible of course that educated people might give rise to unfavourable externalities—intellectuals have a long record of involvement in revolutions! The first point to establish therefore is whether there is a greater likelihood of attaching positive or negative externalities to our brain drain categories.

The most convenient way to approach this question is to view it again in terms of total lifetime patterns. In the case of public services it might be expected that over his total lifetime, a highly qualified person residing in the one country since birth, would on balance make tax payments that were at least equal to his receipts of public services, and at best exceed them. This presupposes that the tax structure is at least proportional to income and at best is progressive, and also excludes the complications posed by any general growth in per capita income levels. If one relaxes this latter assumption, the position is less clear; any continuing growth in income levels could mean that receipts of services by people in their retirement years would exceed any earlier net tax payments, so that they would on balance draw benefits from posterity.

In the case of brain drain emigrants one needs to know how their tax payments and receipts of services will be apportioned as between the home and the immigrant countries. The usual expectation is that up to the completion of their studies such people would have been net recipients of services. Thereafter throughout their working lives they would be regarded as net tax payers. If emigration takes place fairly soon after graduation, the emigrant would on balance be a net recipient of services from his home country. The later the age at which emigration takes place, the less likely this result. Private sector externalities suggest a somewhat different pattern. The work-related external benefits, such as the doctor or research examples quoted above, cannot by definition arise until the education is completed and work begun. If these are the dominant form of externality arising from high level education then emigration would mean that the benefit would accrue to the immigrant area, though the later the age at which emigration occurs the smaller the transfer. It is possible, as noted above, that the externality arising from higher education produces unfavourable effects on the community, but the bulk of the discussions on this question incline to the view that the externalities are favourable and this may be accepted here, since it does not affect the relevant point which is that the externalities do not occur until the post-education period. A subsidiary point of relevance is that the work-related externalities are likely to be of greater value in the middle or later period of a graduate's working life, since it is rarely the case that the younger person is sufficiently experienced or specialised to make major contribut-ions. Put somewhat differently one might say that it would be more likely that the externalities if any associated with younger graduates are likely to be more general in character, and hence could be supplied by other members of the same profession in the event of emigration by some, whereas with an older member there is a greater likelihood of specialised knowledge giving rise to unique externalities which are transferred with him should he emigrate.

On the basis of this type of exposition it is usually assumed that brain drain emigration results in losses to the home country, both through receipts of public services prior to emigration, and the loss of private externalities. This view is not without its critics. Grubel and Scott[3] have contended there is not necessarily any loss of this type. In the case of private externalities they suggest that with older emigrants, where unique losses may occur, many of the results of research work will flow back to the home country as part of the normal international dissemination of knowledge, and that since this emigration is frequently associated with a transfer to a more favourable work environment the total output of such externalities may be increased, with overall gains to the world community, and no necessary loss to the country of origin. With younger graduates, who would constitute the bulk of the movement, they contend that there are short-term transitional losses in training replacements but no permanent losses. With public services, they keep within the bounds of usual discussion in assuming that receipts of public services in general are propor-tional to income. However, with the provision of education they take the view that society is a continuing organism in which publicly-provided education represents an intergenerational transfer from the currently productive generation to the young. These latter will in turn eventually repeat the process, so that the average burden falling on his contemporaries as a result of an emigrant's departure will not be changed by his movement because the emigrant will take along with him not only his contribution to tax revenue in the immigrant area, but also his children, on whom the education component in this revenue will be spent.

[3] H. Grubel and A. Scott, 'The International Flow of Human Capital' *American Economic Review*, Papers and Proceedings (May 1966).

These views, though plausible, are scarcely convincing. In the case of research benefits, Thomas[4] has pointed out that much of the international exchange of knowledge is not free, but is paid for under licence agreements and similar arrangements. The description of the losses associated with young graduates as being short-term or transitional would be of little consolation in a situation where there was continuing emigration (and it was shown earlier that continuing movements could occur), since there would then be a continuing series of such losses. Finally there is the treatment of educational costs, which appears more realistic in that it allows for the fact that parents do finance at least some of their children's education. However, their suggested approach in fact raises a quite separate issue, namely whether the individual or the family should be the unit of measurement for the entire treatment of brain drain effects.

Up to this point it is the individual who has been used as the basis for measurement. If a family unit is adopted in order to allow for payments between parents and children in respect of education, it seems logical to apply this approach for measuring all other payments and benefits. The first question then becomes the definition of the family—specifically the determination of the points at which membership of one family terminates and that of a new one begins. It is not however necessary to resolve this problem here. The second problem is to unravel the complications of intra-family but international transactions. Thus the child who emigrates and then sends money home to his parents may be making an internal family transfer (an offsetting payment to the receipts of the child by way of education and other services when he was younger); but it will appear statistically as an international transfer. The family is not then a particularly convenient unit to use for any calculations. More important is that it is a clumsy basis for discussing the issue in question. The conclusion drawn by Grubel and Scott only applies to situations where income and educational levels are constant over time. In cases where there is income growth, and/or an increase in the proportion of young people receiving education the payments of the parents could not reasonably be regarded as meeting the expenditure on their children's education. The more consistent and feasible approach

---

[4] B. Thomas, 'The International Circulation of Human Capital', *Minerva,* summer 1967.

therefore is to use the individual as the basis for all calculations, in which case one can regard the migration of people who have received publicly-financed education as giving rise to an externality.

THE UNEMPLOYMENT CASE

The two previous cases have presupposed full employment, so that migration was either influenced by non-economic factors, or by better economic prospects in the immigrant area. In both instances one might expect the 'pull' from the immigrant area to be the dominant influence. The next case is where migration is due, at least in part, to the 'push' from the home country because of a lack of employment opportunities. For convenience one may distinguish two forms of unemployment situation: that where the unemployment is of a temporary nature associated with fluctuations in demand and supply conditions, and the second where unemployment is of a more continuing long-term nature presumably because of some structual imbalances in the economy.

The presence of the first type, short-term unemployment, does not appear to lead to any substantial extensions or alterations in the earlier analyses, although it does result in some smaller additions. One purpose which it serves is to provide a plausible motive for the type of equilibrating migration considered earlier, since unemployment creates a major need for income on the part of those affected, and hence enhances the attractiveness of other areas where work is available. This pressure to move is reduced if unemployment benefits or other forms of income maintenance are available in the home area. The actual payment of such benefits represent a cost to the area making them, and so there is a specific short-term gain to the emigrant area if migration takes place, since the need for these payments is then avoided. This does not however mean that there is an overall benefit to the emigrant area. Since it is assumed that the unemployment is only temporary, the subsequent life-time pattern of costs and benefits may entail net losses of private transfers and tax payments to the home area of a type described in the preceding cases. Hence it could be worthwhile on economic grounds to incur the short-term costs of unemployment benefits in order to secure the longer life-time gains of retaining an individual at home. Viewed in more aggregative termst here is also the consideration, however, that migration can serve to speed up the process of restoring full employment, which it does not only by reducing the numbers of unemployed, but also by providing a potential stimulus to export demand, since it might be expected that the new emigrants would be interested in purchasing the products of their home country. Any such increase in exports should help to boost output and hence restore full employment.

It is also relevant to enquire what would be the likely composition of migration due to unemployment. The data on unemployment indicate that the incidence of unemployment is lower among the more highly qualified groups. In part this is associated with the wider range of jobs open to such people, one example of the several 'option benefits' which Weisbrod[5] attributes to education. It would be expected on this basis that there would be a lower than average proportion of brain drain migration associated with unemployment. It is also possible, however, that the lower unemployment level among the highly qualified would conceal a situation where many of them were occupying jobs at a lower level than those for which they were qualified, so that they might face strong incentives to emigrate.

A further possibility is that the observed lower incidence of unemployment among highly qualified personnel is associated with the higher general incidence of migration among this group. One way to check whether this higher general emigration was itself a major cause of lower unemployment among the highly qualified, would be to compare the incidence of such unemployment in areas of emigration with the incidence in immigrant areas.

With the long-term or structural form of unemployment there is a somewhat different picture. Now when the individual emigrates he experiences a substantial gain in lifetime income by comparison with a position of prolonged unemployment at home. Likewise the effects on the rest of the home economy are altered. No longer is it reasonable to regard emigration as leading to the loss of a series of tax receipts, or other benefits; in fact remaining at home would entail payment of a continued series of unemployment and related benefits. It might seem at first glance that it would be economically rational in such circumstances to

[5] B. Weisbrod, *External Benefits of Public Education*, Princeton 1964.

pay the emigrant's fare rather than pay him to stay, especially since there is the prospect of some return benefits both by way of gifts and through the stimulus to exports.

However, it is important to examine the implication of longer term migration more fully before reaching any such conclusions. First it is relevant to ask how structural unemployment occurs. The normal answer would attribute it to differences in the composition of output and/or the production method used. These differences would in turn be related to technical progress. Several influences would in turn affect the level of such progress, but to successfully sustain a satisfactory rate of progress calls for an adequate rate of capital accumulation—both physical in the form of equipment, and human in the form of skilled personnel. Migration clearly alters the distribution of human and physical capital as between areas, so the form and likely effect of these changes must be explored.

First, migration by altering population densities either produces a better balance between human population and land or other natural resources in which case it is presumably a good effect, or it alters this distribution adversely, by creating or accentuating imbalances. Secondly, migration alters the population structure so that the immigrant area has a greater proportion, and the emigrant country a lesser proportion, of working population to dependants. Effects of this latter type are likely to result in an intensification of any initial differences since a higher dependency ratio makes it more difficult to provide a living standard comparable to that of the full-employment area.

The brain drain component in such emigration is less clear. It could be expected that the home country would have a smaller proportion of highly qualified personnel in its labour force than the full employment area. This in turn would call for a relatively smaller supply. The actual supply will depend in part on the capacity of the country to sustain education from its relatively lower income level. There may also however be a strong interest in and pressure for education because, if there is structural unemployment and hence an extensive need for migration, education will be valued for its capacity to facilitate such movement. Hence it might well be the case that there is a substantial excess supply of educated people, and consequently a substantial brain drain component in the migration. Where this is so it will further intensify any tendency for the initial disparities between the full-employment and home country to become permanent, because it will accelerate the rate of (human) capital accumulation in the immigrant area, while the effort of providing this capital investment will tend to depress the level of accumulation in the home area.

Consideration of the longer-term implications may suggest therefore that it would not be economically correct to assume that the migration of unemployed will represent a gain to the emigrant area. This is not to preclude the possibility that it would be so desirable in some cases. The contention is rather that no general answer is possible.

THE IRISH CASE

The cases discussed in the preceding section describes characteristics which would apply to many countries. The predominant tendency in brain drain movements of recent years has been one of a net migration from the poorer, less developed countries, and regions, to the wealthier zones. The Irish pattern of movement can be viewed as part of this trend, since the bulk of Irish emigration is on balance to higher-income countries such as Britain and the U.S. By taking account of the specific circumstances affecting the Irish community, it is possible to identify more fully the extent to which the various general characteristics would apply in the Irish case. Thus it would be expected that given the facility of a common language, the short distances, the awareness of British conditions gained from newspapers and other media, the large volume of trade and the freedom of movement between the two countries (no passports are required) that there is less reason to expect any pronounced emphasis on brain drain movements between Britain and Ireland since the obstacles to movements by other groups are much weaker than they would be for other emigrant/immigrant countries.

Despite such individual characteristics, it may nonetheless be presumed that any assessment of the gains or losses accruing to the rest of the Irish economy from brain drain movements would follow the lines indicated by the general cases of the previous section. Thus data would be needed on the expected pattern of production and consumption associated with highly qualified Irish personnel in both Ireland and the countries to which they emigrate. This would enable estimates to be made of the likely transfers by way of wealth movements and gifts which would result from migration. In addition data would be needed on government tax and expenditure patterns in the relevant countries, in order to estimate the extent to which externalities of this type would arise. It would further be desirable to estimate any private externalities and finally, given the unemployment position in Ireland, it would be important to identify the impacts on exports, and rates of capital accumulation.

Assuming that such data were available it would enable one part of the answer to the question of whether or not brain drain constitutes an economic loss to be determined, in that it would indicate whether a highly qualified person was more valuable to the rest of the community at home or abroad. This however is only part of the data which is needed for policy-relevant purposes. It is also necessary to calculate the net benefits or costs associated with second level school graduates—those who remain at home and those who emigrate. By comparing this data with that for the brain drain categories it would be possible to establish the difference in net benefits or costs between the two groups. It could then be seen whether it was economically preferable to have people emigrate as highly-qualified personnel or with some lower level of qualifications.

It should be emphasised that calculations of this nature are predominantly marginalist, in that they seek to estimate the effects which on average may be expected to stem from the movement of an individual or small group of people. It would also be necessary to place these partial equilibrium type calculations into a more general setting before reaching any overall economic judgement. An illustration of such an overall requirement was the example given in the unemployment case discussed above, of the effect which differences in the aggregate rate of capital accumulation (in this instance human capital) might have on a country's rate of economic growth.

A second example of such general effects and one which could be relevant in Irish circumstances, concerns the impact which educational policies cum brain drain movements might have on income relativities and absolute levels of pay. In a small economy it would be expected that various shortages and bottlenecks in the supply of skilled and highly qualified personnel would arise in the course of development. Such shortages coupled with the comparatively small numbers involved (viewed as a proportion of the total labour force) could be expected to result in a rise in their pay rates, relative to those of other groups. In a fully-functioning, frictionless market economy, these changes in relativities could be expected to proceed smoothly. In reality, any such movement may meet with resistance and create problems. First, any change in relativities which does occur will mean rises in the pay rates of the favoured groups, rather than reductions elsewhere, since it is virtually impossible to bring about a lowering of pay rates. Secondly, any attempt to alter relativities, especially in favour of higher paid groups, normally meets with strong opposition, so that pressure for all round pay rises typically follows any initial increases for groups in short supply. The consequences of upward movements in general pay rises will be to generate price increases, and this inflation could produce substantial adverse effects on international competitiveness, which in the case of a small country would have a significant impact on its development, since external trade is of greater importance for such a country.

One possibility in seeking to minimise problems of this type would be to have an educational/manpower policy which aimed at providing adequate supplies of highly skilled personnel. Since the training period for these categories can be comparatively lengthy, this means that many decisions must be based on esti-mates of the future demand and supply position. To allow for the inevitable inaccuracy of forecasts, a successful policy would have to err on the side of oversupply rather than undersupply. The main possibilities for absorbing this surplus are threefold: there could be a faster than planned or expected expansion in employ-ment of such people, with presumably a tendency for their wage levels to decline (or rise less rapidly) in relative terms; secondly, there could be a tendency for the surplus to be absorbed in lower skilled (and presumable lower paid) jobs; or thirdly they may emigrate, constituting a brain drain.

This last result could arise in situations where on the basis of an individual calculation there is a loss attached to each brain drain emigrant. Such economic losses would impair the pace of development, yet they may constitute a smaller impediment than an alternative where surpluses were eliminated, but where skill bottlenecks were a source of general inflationary pressures. It is difficult to envisage how any adequate empirical testing of this point could be made since no direct comparisons of the alternative situations posited are possible. It would however be feasible to build up circumstantial evidence over a period of years, by analysing rates of growth in occupational categories, rates of change in pay, and rates of change in educational qualifications of occupational groups.

This reference to the possible merits of a brain drain as an anti-inflationary mechanism would arise only where the market system was not functioning in a fully perfect manner. In this instance it was assumed for example that pay relativities were influenced by non-economic considerations such as social attitudes or equity concepts.

There are other possible factors of a non-economic nature which might be relevant in the specification of any detailed model for Irish conditions. Traditionally, for example, a significant fraction of Irish emigrants went to the U.S. The changes in U.S. immigration laws since 1965 have meant in practice that it is easier for brain drain personnel to gain access, and have reduced the total number of migrants. Such changes in the total and composition of migration between the two countries could have the effect of shifting the emphasis from a less costly to a more costly pattern from the Irish viewpoint, but again, as with some of the earlier examples, it could still be the case that the more costly type was still an economically rational response in the changed circumstances.

It is not proposed, however, to explore any such constraints or obstacles any further here. The intention was rather to illustrate the point that the results of any generally applicable economic analysis would need to be related to the specific conditions prevailing in an economy before reaching any policy-oriented conclusions.

SPECIFIC ISSUES

The logical extension to the foregoing analysis would be to evaluate empirically the relative size and influence of the major factors involved in Irish brain drain movements. An examination of the available material suggested that no satisfactory results could be obtained from these sources. This concluding section is accordingly concerned with the more limited task of commenting on some of the issues arising from the paper by Professor Lynn.

The first question discussed by Lynn is the size of the brain drain movement, which he deals with in terms of the fraction of newly graduating students who emigrate. The actual figures obtained need not be discussed since they have received considerable comment; it may be accepted that emigration in the relevant groups is large. Given the earlier discussion, the more interesting question is whether brain drain emigration is on a larger scale than that of other groups.

One way of seeking to answer this question has been used by Grubel and Scott (in the immigration study noted above). They calculated the number of brain drain migrants as a fraction of the total number of brain drain personnel and compared this with the total number of migrants from the total population. A ratio of 1:1 would mean that brain drain migration was on a par with general migration, a figure greater than one would mean that it was above average. For Ireland their data yielded a ratio of 14:1. It should be noted that this referred to immigration into the U.S. by Irish people, and not to total emigration from Ireland, that it dealt with scientists and engineers, and that it referred to the latter part of the 1950s.

It is not feasible to apply this form of calculation to the overall Irish situation because there is no data on the educational qualifications of the population. This position will change shortly, because the 1971 population census will collect such data for the first time. The availability of this information will, however, raise the question of the most appropriate way in which to measure relative emigration. The method of comparison used by Grubel and Scott is of a cross-section kind, that is comparing the experience of different groups over the same period of time. There is the alternative of comparing the experience of the same group over different periods of time. This could be done by following the experience of given cohorts of children from birth onwards, establishing the extent to which there were differences in the lifetime emigration patterns of children receiving varying amounts of education. Cohort analyses of this type would be a useful supplement to the cross-section data because the latter reflect the many short-term influences which go to produce the emigration of any one time period.

The second issue discussed by Lynn is the cost of the brain drain. Several methods of valuation and type of cost are discussed by him, not all of which are of an economic nature. The first of the economic forms of calculation treats the cost of the brain drain in terms of the cost involved in providing the necessary university education.

The second cost calculation is based on the potential tax losses associated with brain drain migrants. Drawing on British data which estimates the lifetime tax losses associated with a management migrant at £23,000, Lynn obtains a figure of £46 million as the annual loss arising from graduate emigration of 2,000 people.

Neither of these measures can be accepted as satisfactory. The first tells us the cost of educating graduates, nothing else. It gives no indication of the gains or losses associated with the subsequent use of these graduates in different locations. The second measure is more relevant, but is unsatisfactory on at least two grounds. One is that loss of tax revenues tells only part of the story. It is also necessary to offset against these the savings on government services which no longer have to be provided. It is true that the British source from which Lynn draws his data advances the view that the higher-paid do not derive many benefits from welfare services, and that other services such as defence are ones where the level of spending is unlikely to vary in response to small migrational movements. It is interesting to contrast this British analysis (designed to show that there is a loss from brain drain movements) with the Grubel and Scott analyses quoted earlier (which contend that there is not any significant loss from the brain drain). It will be recalled that the American authors assumed that receipts of government services were broadly in proportion to the tax payments made. In the absence of any detailed study as to how services are in fact distributed among the population it seems reasonable to mark both cases as not proven.

The second objection to using loss of tax receipts as a measure of brain drain costs in Irish conditions is that it assumes graduates could in fact obtain employment in Ireland. Lynn refers to this aspect briefly, and suggests that there is no reason why employment could not increase considerably. The short answer to this is that employment has not grown at the necessary rate—for whatever reasons. The short-term and longer-term consequences of migration in conditions of unemployment have already been dealt with in the preceding section. One need simply state that in such circumstances brain drain movements could constitute either a gain or loss, depending on precise conditions.

The remaining sections of Lynn's paper need not be discussed since they deal with possible methods of curbing the brain drain. It may be worth noting, however, that this discussion proceeds on the basis that the existence of a loss from brain drain is a sufficient reason for seeking to eliminate these losses. Again from the earlier analysis it may be contended that this is insufficient since the loss, if any, from brain drain may be smaller than the losses which would arise with any feasible alternatives.

The conclusions to be drawn from this examination of the brain drain issue are several. First, brain drain movements can produce substantial economic effects on the countries concerned. Second, it is not possible to say on a priori grounds whether a brain drain results in net losses or gains to the emigrant country; there is however a stronger presumption that the immigrant area gains. Thirdly, it has been established that a considerable amount of data would be needed to produce satisfactory empirical results for any country. Fourthly, no such data exist for Ireland. These conclusions imply that policy decisions in this area must either be based on non-economic considerations. or that economic factors can only be considered in the most general terms. This situation can be expected to alter significantly in the coming years, as more extended data become available, which will enable the relevant economic questions to be answered.

CHAPTER VIII

Wage Inflation and Wage Leadership in Ireland 1954-69

C. MULVEY and J. TREVITHICK

INTRODUCTION

THIS chapter is a preliminary investigation of the role which a 'wage leader' may play in the process by which average money earnings change. In particular the chapter examines the characteristics of the group within the labour force which may act as a wage leader, the conditions under which such a group will tend to function as a wage leader and the manner in which a leading wage settlement will be transmitted into the general wage level. All of the data used in this chapter relates to the Republic of Ireland, largely because of the smallness and compactness of its industrial labour force and the relative simplicity of its wage determination system, both of which allow a fairly straightforward investigation of phenomena which in larger, more developed countries may be obscured by the complexity of the process of wage determination. Hence although the data and analysis apply particularly to the Republic of Ireland, there may well be significant implications arising out of the findings for other economies, in particular the U. K. and the U.S.A.

Previous studies of the determinants of the rate of change of money wages in Ireland have produced findings broadly in line with similar studies carried out for the U.K. and the U.S.A. O'Herlihy[1] found that the level, and the rate of change of the level, of non-agricultural unemployment and changes in retail prices were significant determinants of the rate of change of average industrial hourly earnings. Cowling[2] found that changes

[1] C. St. J. O'Herlihy, *A Statistical Study of Wages, Prices and Employment in the Irish Manufacturing Sector,* Economic Research Institute, Paper No 29, (Dublin 1966).

[2] K. Cowling, *Determinants of Wage Inflation in Ireland,* Economic Research Institute, Paper No 31 (Dublin 1966).

in retail prices, the rate of change of profits and the rate of unionisation and the rate of change of unionisation were significant explanatory variables in determining the rate of change of money wage rates and earnings. These findings fit the general pattern of results obtained for U.K. and U.S. data in the post-war period. However, neither study is entirely satisfactory since both largely ignore certain important institutional influences on wage determination in Ireland, and, moreover, the O'Herlihy study appears to lose a significant part of its explanatory power after 1959. In addition there is a certain conflict between the findings of the two studies and neither possesses a particularly effective explanatory power.[3] The starting point in this chapter is therefore an examination of the institutional factors which influence the rate of change of hourly earnings and an hypothesis is derived from that examination which is tested in the second part of the chapter.

THE BACKGROUND

The process of wage determination in Ireland since 1946 has taken the form of a series of wage rounds some of which have been the subject of national agreements between the Federated Union of Employers (F.U.E.) and a trades union Congress, and others which have been more or less free-for-alls. However, both types of wage round have been characterised by a high degree of uniformity in the magnitude of wage settlements obtained throughout the labour force and by a relatively well defined time span. David O'Mahony described these wage rounds in detail[4]. To date there have been twelve complete rounds since 1946 and the 13th round is likely to begin in late 1970 or early 1971. Of these wage rounds six have been the result of a national agreement and the other six have been more or less spontaneous. However, since 1958 there have been six rounds and only one was subject to national agreement.

A characteristic of these wage rounds has been a high degree of uniformity in the magnitude of the settlements made throughout the industrial labour force. In the case of rounds which follow a

[3] See Appendix 4.
[4] D. O'Mahony, *Economic Aspects of Industrial Relations*, Economic Research Institute Paper No. 24 (Dublin 1964).

national Congress/F.U.E. agreement this was inevitable since such agreements laid down either a minimum or maximum guideline in either percentage or flat rate terms. Individual bargains struck within the context of such an agreement invariably applied the terms of the agreement directly. The result of this process has been to interfere little with the relative structure of occupational or industrial earnings. However, one might expect those wage rounds which have not been subject to the terms of a national agreement to produce a varied pattern of individual settlements. Indeed, one might intuitively expect such rounds to incorporate a variety of different settlements designed to adjust the wage structure in order to correct imbalances in particular labour markets caused by the rigidity of settlements applied in relation to national agreements. However, this has not happened in practice.

Much the same tendency towards uniform wage settlements throughout the industrial labour force may be observed in wage rounds which have not been subject to national agreement. The reason for this tendency is the paramount importance of the comparability criterion espoused by Irish trade unions. The widespread acceptance of this criterion by Irish trade unions (and many Irish employers) derives partly from the structural form of the Irish trade union movement but has also been encouraged by the implied endorsement of such a policy by the national wage agreements and by the activities of the Labour Court. Indeed one of the most significant institutional forces in maintaining stability in the structure of wages in Ireland has been the Labour Court. In any case a very large area of the collective bargaining system in Ireland operates solely on the basis of comparability as a criterion of wage rate change. In this respect Ireland is not unlike either the U.K. or the U.S. except perhaps in the degree of importance which attaches to comparability and the extent to which it is utilised.

Given then that the primary criterion applied by trade unions and employers in collective bargaining is comparability with similar types of labour employed elsewhere, a very high degree of stability in the relative structure of wages is to be expected in the short run. Over the longer period there has been a tendency for wage differentials to narrow somewhat, although this process appears to have been halted or even reversed around the early 1960s, and for some changes in inter-industry differentials to emerge. The former tendency has been gradual and probably reflects changes in emphasis in trade union wage policy in response to changing patterns of unionisation. In any case, the trend towards a narrowing of differentials appears to have taken place with the consent of skilled workers, who have been party to national wage agreements which have yielded flat rate wage increases, and appears to have ceased when skilled workers withdrew this consent.[5] The changes which have occurred in the inter-industry wage structure are almost entirely reflections of changes in the structure of employment.[6] Hence, even over the longer period, the structure of wage differentials and relativities has tended towards stability with trade unions and workers themselves initiating whatever adjustments in the structure have been considered necessary or desirable. The main point here is that the criteria of collective bargaining in Ireland, as applied by both trade unions and employers, ensure that when any adjustment occurs in the relative structure of wages it will be quickly generalised throughout the wage structure in such a way as to restore the *status quo ante*. Further, it may be the case that when some such adjustment in the wage structure is threatened (e.g. by a claim submitted but not yet settled) the trade unions and employers may supply the machinery for its rapid transmission throughout the wage structure by making a national agreement. Such a procedure has the attraction of ensuring orderly and strike free progress towards a general adjustment in the wage level which would approximate to that which would be achieved through individual bargaining in any case, provided that general agreement on the order of magnitude of the settlement exists.

The critical question which obviously arises in this context concerns the origin and determinants of the initial disturbance (or threatened disturbance) in the wage structure which sparks off the general adjustment in the wage level. The idea of a wage leader is clearly a possibility here. If some group of workers were continuously in a favourable bargaining position within the economy then one might expect that group would tend to set

[5] R. Roberts, 'Trade Union Organisation in Ireland', *Journal of the Statistical and Social Inquiry Society of Ireland*, XX, II, (1958–59).

[6] K. Kennedy 'Growth of Labour Productivity in Irish Manufacturing', *Journal of the Statistical and Social Inquiry Society of Ireland*', XXII, I, (1968–69).

the pace for the rest of the unionised labour force by periodically striking key bargains which, via the comparability and relativity criteria, would set the whole structure of wages in motion in such a way that the general level of wages would adjust upwards in line with the terms of the key bargain. If such a group could be identified and the influences which determine the timing and magnitude of the key claim or bargain discovered, then the determinants of the rate of change in average industrial earnings would then be apparent.

A number of possible potential wage leaders within the labour force were investigated. The notion of an *industry* wage leader, such as was identified by Eckstein and Wilson[7], although intuitively appealing, proved irrelevant to the Irish context, although two industrial groups, building and construction and electrical contracting, clearly did play a leadership role on a number of occasions. The possibility of an *occupational* wage leadership group was then examined and this proved a more fruitful approach In every wage round in which it is possible to distinguish the group striking the key bargain it was noted that one occupational category of skilled labour was invariably represented in the key group. This was the category of electricians and electrical fitters. On occasions this occupation struck a key bargain *on an occupational basis* e.g. the electrical contract shops agreement of 1968. On other occasions the occupation was only an influence on a key *industrial bargain* e.g. in the building and construction industry in 1967. On still other occasions (more recently) it has been a substantial element in a key bargain of a *skill* type e.g. the maintenance craftsmen's agreement of 1966. This occupation is the *only* common element which can be identified in every key bargain during the 1960s. This constitutes *prima facie* evidence that if indeed any single occupational group within the labour force consistently played the role of wage leader then it has probably been electricians.

The most striking economic characteristic of the labour market for electricians is that the unemployment rate has continuously remained at low levels throughout the post war period and has remained at extraordinarily low levels during the 1960s. The estimates of unemployment rates amongst electricians (see

[7] O. Eckstein and T. A. Wilson, 'The Determination of Money Wages Rates in American Industry', *Quarterly Journal of Economics,* Volume 76, (1962).

Appendix 2) indicated that during the 1950s unemployment varied between an annual average high of 2·68 per cent and an annual average low of 1·57 per cent. Over the same period aggregate non-agricultural unemployment varied between an annual average high of 9·6 per cent and an annual average low of 6·8 per cent. During the 1960s unemployment amongst electricians varied between an annual average high of 1·28 per cent and an annual average low of 0·81 per cent. During the same period aggregate unemployment varied between an annual average high of 6·7 per cent and an annual average low of 5·6 per cent.

Data problems make it impossible to make detailed comparisons of unemployment rates in different occupational labour markets. Data problems also preclude the construction of an index of excess demand for any occupational labour markets. However, there is some evidence[8] and the results of surveys carried out by the Department of Labour, (unpublished) tend to confirm, that:

(a) there has been continuous excess demand for electricians since at least 1961;

(b) since 1961 the level of excess demand for electricians has been greater than in any other occupational labour market;

(c) there were severe inequalities in the incidence of unemployment between different occupational labour markets during the 1960s; and

(d) in general, unemployment is greatest in the markets for unskilled labour and lowest in the markets for skilled labour.

This would indicate that there is a degree of structural unemployment in the Irish labour market despite the relatively high overall unemployment rate and that this situation prevailed throughout the 1960s. Despite this high overall rate of unemployment wage inflation in Ireland proceeded during the 1960s at a rapid pace, indeed in recent years more rapidly than in any other country in

[8] R. C. Geary and J. G. Hughes, 'Certain aspects of Non-Agricultural unemployment in Ireland', *Economic and Social Research Institute* Paper 52 (Dublin 1970).

Europe. One way of explaining this phenomenon is to assume that it is the result of the uneven distribution of unemployment among sectoral, occupational or industrial labour markets.[9] This hypothesis predicts that the general level of wages can rise, despite a net excess supply of labour is aggregate, if there are inequalities in the distribution of unemployment between the different labour markets. The hypothesis however depends on differential rates of change of wages in different labour markets leading to a positive rate of wage inflation on average under certain conditions. In the Irish case however, the fact is that wages have tended to increase at fairly uniform rates throughout each sectoral, occupational and industrial labour market. What appears to be a more plausible explanation of this phenomenon is that the economic conditions obtaining in a key occupational labour market will tend to be reflected, as a result of the application of various comparability criteria by trade unions, in the rate of change of the *general* level of wages.

So far there is no more than *prima facie* evidence to suggest that if indeed a wage leader does exist in the Irish economy then it is probably electricians in one or other of the bargaining units of which they comprise a significant element. In the next section of this chapter a hypothesis is proposed which would account for the leading role of electricians in the Irish wage determination system and in the following section the predictions of the hypothesis are tested.

THE MODEL

HYPOTHESIS

The demand for labour, being a derived demand, may be supposed to vary directly with the demand for the output of the firms employing that labour in the short run. More specifically, the demand for electricians may be supposed to vary directly with the demand for the output of those firms employing electricians. It is assumed that in the short run such a variation will be a proportional variation, although this assumption is not essential to the hypothesis. The demand for labour in the short

---

[9] See R. G. Lipsey 'The Relation between Unemployment and the Rate of Change of Money Rates in the UK, 1862–1957: A further analysis', *Econometrica*, Volume 27, 1960.

run will therefore be assumed to be a constant proportion of the demand for final output. This can be written symbolically:

$$L^d = \frac{1}{u} Y^d \quad \ldots\ldots\ldots\ldots\ldots(1)$$

where L^d is the demand for any category of labour (in the model, the demand for electricians) which finds its origin in the need to meet current production requirements; Y^d is the demand for the output of firms employing this category of labour; and u is the output/labour coefficient.

Equation (1) may represent the demand for a category of labour under certain labour market conditions but will certainly be incomplete in others. In expanding this point it may be convenient to think of the category of labour as representing electricians as an occupational group. When electricians are in excess supply, there will be little pressure on employers to forecast future requirements for electricians since any increase in Y^d will

simply increase the demand for electricians by a proportion $\frac{1}{u}$, and

this extra demand can be met simply by hiring more electricians at the going market rate. In this case equation (1) will accurately reflect the total demand for electricians at any time t. However, when the demand for electricians is very high, and has been continuously high over a period of years, as has been the case in Ireland during the 1960s, there will be a certain pressure on employers to attempt to predict future changes in demand and to hire and dismiss labour in anticipation of such changes. When electricians are in short supply continuously, employers will have learned from past experience that as aggregate demand increases so the number of electricians available for hire diminishes until a point is reached at which none is obtainable. Past experience will therefore suggest to employers that there are costs associated with failure to predict future requirements for electricians, and to hire accordingly, since bottle-necks will appear in the production process and these will prevent employers from fully satisfying the extra demand for their product. It is clear that the cost of failing to anticipate future labour requirements when the labour market is tight will vary directly with the degree of excess demand in each occupational labour market. Since it is observed that the level of excess demand in the labour market for electricians is greater than that existing in any other occupational labour market, the cost of failing to anticipate future requirements for electricians will be greater than in any other occupational market.

The question now arises as to how employers predict future labour requirements. The simplest assumption is that employers will assume that demand will increase in period $t+1$ by a constant proportion of the change in period t. In this case the hiring or dismissal of electricians will depend upon the change in demand between period $t-1$ and period t. The most immediate behaviour of demand is thus assumed to form the basis for expected changes in demand. Consider the general case of an individual labour market in the short run.

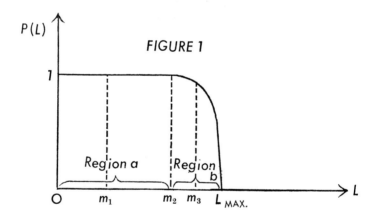

When labour markets are slack, entrepreneurs will know that, subject to random fluctuations, they will be able to hire x men if x men are required. Call the number of extra men required $x^\star$ and the number of men actually hired in response to this requirement x. (Throughout we will be speaking of hiring extra men, although similar arguments apply, *mutatis mutandis*, to the dismissal of men.) In slack labour markets, the proportion $\frac{x}{x^\star}$ will be approximately unity; entrepreneurs will be confident that if any future changes in demand for their products occur, they will be able to find the extra labour almost immediately. Hence, the expected value of the distribution of $\frac{x}{x^\star}$ is almost unity. Thus, if L is in the interval $0<L<m_2$ (i.e. in region a in Fig. 1), no anticipatory hiring need occur since future demand increases can be met without any difficulty (in the labour market at least). Call the expected value of $f\left(\frac{x}{x^\star}\right)$ the probability of being able to find one extra man if one extra man is required and call it $P(L)$. In region a (the interval $0<L<m_2$), $P(L)=1$. In Figure 1 $P(L)$ is the y-axis. Region a in Figure 1 is defined as that domain of L in which $P(L)=1$ and $P'(L)=0$.

In region b, the expected value of the distribution of $\frac{x}{x^\star}$ is distinctly less than unity.

THE COSTS OF FAILING TO ANTICIPATE LABOUR SHORTAGES.

Suppose the number of men required increases by an amount z. The mathematically expected number of men actually hired will be

$$(1)\ \ldots \ \int_p^q P(L)dL \text{ where } q-p=z.$$

If both p and q fall in region a, the integral in expression (1) will be equal to z since $P(L)=1$ in this region.
On the other hand, if q alone or both p and q fall in region b, the integral expression (1) will have a value less than z so that the mathematically expected number of men actually hired will be less than the number required,

i.e. $$\int_p^q P(L)dL<z$$

For example, if in region a, $p=m_1$ and $q=m_2$, the mathematically expected number of men hired will be:

$$\int_{m_1}^{m_2} P(L)dL=z.$$

But if in region b, $p = m_2$ and $q = m_3$, the mathematically expected number of men hired will be:

$$\int_{m_2}^{m_3} P(L)dL < z, \text{ since } P'(L) < 0.$$

The quantity $z - \int_{m_2}^{m_3} P(L)dL$ will be the amount of extra labour demand unsatisfied. The total labour shortage will be $(m_o + z)$

$-\left(\int_0^{m_1} P(L)dL\right)$. m_o is the initial labour requirement, which may or may not have been satisfied and m_i is any point on the L-axis which may lie in either region a or region b.

In the short run it is valid to assume fixed coefficients of production i.e. $Y^d = uL^d$ and $Y = uL$ where L^d is the total required labour force, L is the mathematically expected actual labour force, Y^d is the demand for final output and Y is the maximum supply of output.

The expected shortfall in final supply will thus bear a linear relationship to the expected shortfall in labour supply.

$$Y^d - Y = u\left((m_0 + z) - \int_0^{m_i} P(L)dL\right) \text{ where } z = m_i - m_0$$

If $m_1 \leqslant m_2$, $\int_0^{m_i} P(L)dL = (m_0 + z)$ so that $Y^d = Y$ and $Y^d - Y = 0$.

i.e. the expected shortfall in final demand is zero in region a.

If $m_i > m_2$, $\int_0^{m_i} P(L)dL < (m_0 + z)$ so that $Y^d - Y > 0$

The greater is m_i, the greater the expected deficiency in final supply, $Y^d - Y$.

This can be shown diagrammatically in Figure 2.

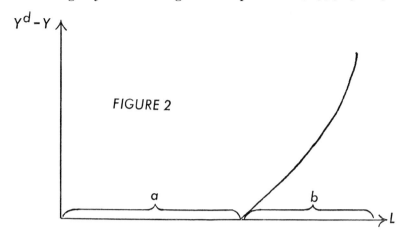

FIGURE 2

When $Y^d > Y$, i.e. when we reach region b, entrepreneurs will commence hiring electricians in anticipation of a shortage in the supply of this occupational group. It is postulated that the labour market for electricians was initially in neo-classical equilibrium with the real wage rate (w) equal to the long-run marginal physical product of labour (u), the access of optimistic expectations will induce entrepreneurs to hire electricians in anticipation of a future shortage of this category of labour. (Following the neo-classical tradition u, the output/labour ratio, is assumed to be fixed in the short-run but variable in the long-run). The number of electricians affected by such anticipatory hiring will be a function of the difference between Y^d and Y. Moreover this anticipatory hiring will force up the money wage rate of electricians, the rate of increase being directly related to $Y^d - Y$. Since it is central to our hypothesis that the wage bargain struck by electricians will set the pace for the rest of the economy, and since most econometric studies confirm the paramount importance of aggregate money wages in determining the general price level, in the long-run the real wage rate for electricians will not be appreciably affected by demand considerations. The real wage rate for electricians will only rise if the aggregate productivity of labour rises; if aggregate labour productivity grows at a rate m, the real rate for electricians will follow a path approximately similar to $w_0 e^{mt}$. The above analysis is independent of the choice of time period; clearly for the purposes of our hypothesis the relevant time period is that period over which entrepreneurs make forecasts of future labour requirements.

As was mentioned earlier, the expected level of demand in period $t+1$ is simply the sum of the observed level of demand in period t and the change in demand between t and period $t-1$. That is to say, $Y^d_{t+1} = Y^d_t + a(Y^d_t - Y^d_{t-1})$. In other words, the expected level of demand in any period $t+1$ depends not only upon the level of demand in period t but also the rate of change of demand. Considering the specific labour market for electricians, the total demand for labour may be written:

$$L^e_{t+1} = uY^e_t + k(Y^e_t - Y^e_{t-1})$$

where L^e_{t+1} is the total anticipated demand for electricians and where $k = au$. Retaining this simple linear form of the relation and assuming that variations in the labour supply are only minor in the short run, an expression for the excess demand for electricians can be written in which not only the rate of unemployment amongst electricians, but also the rate of change of this rate, plays a significant role.

$D^e_t = cU^e_t + d\Delta U^e_t$ where D^e_t is an index of the excess demand for electricians. The restrictions on this equation are that $d = 0$ where U^e is large and $d > 0$ where U^e is small. At low levels of $U^e, \Delta U^e$ will be the more powerful indicator of the level of excess demand since U^e will be close to its floor level.

The familiar relation that the rate of change of the money wage rates of electricians is a function of the level of excess demand for their services is:

$$\Delta W^e = f(D^e).$$

But earlier in this paper it was argued that the rate of change of average wage rates will be a direct function of the rate of change of the leading occupational wage rate, i.e. that:

$$\Delta W = j(\Delta W^e)$$

Postulate [10] that the rate of change of average hourly earnings in

---

[10] By using the relation described to test this hypothesis it is implied that wage rates determine the level of earnings. That is, earnings 'float' on top of wage rates and wage bargains do not merely represent consolidation of previously gained increases in earnings. It was impossible to go into this problem in detail but a test was made of the relation between changes in hourly earnings and

transportable goods industries is a function of the rate of change of average hourly wage rates:

$$\Delta E = z(\Delta W)$$

Hence:
$$\Delta E_t = h(D_t^e)$$

This last relationship provides a testable hypothesis and the findings are set out in the following section.

In this section the rate of change of money wage rates of electricians has been assumed to be a function of the level of unemployment for their services, but this relation will only hold for relatively high levels of unemployment (low levels of excess demand). The theoretical underpinning for our observation is provided in that when the absolute level of unemployment is low among electricians, the rate of change of unemployment among electricians explains more of the variation in the rate of change

---

changes in hourly wage rates which indicates a high degree of association between the two variables in Ireland for the period 1945–1969.
$$\Delta E_t = 2·28 + 0·75\Delta W_t + e_t \; ; R^2 = 0·83, \quad \text{standard error in parentheses.}$$
$$(0·10)$$
Hence, even in years when no wage round occurred (which constituted the majority of the observations) there would appear to be a high correlation between the rate of change of average hourly earnings and the rate of change of hourly wage rates. That is, in years in which hourly wage rates increaseed only slightly (inter-round years), average hourly earnings varied very much as predicted by the above equation. For this reason the consolidation effect exerts probably only a minor influence in the process by which earnings increase in Ireland and that the prime mover is hourly wage rates. This view corresponds with the findings of Dicks-Mireaux and Shepherd[11] and Hines[12] using U.K. data and is in line with '. . . the (entirely commonsense) view that national negotiations between trade unions and employers associations are not irrelevant, but affect the level of earnings.'[13] It is also of course a view which corresponds closely to our analysis of the general nature of the process by which the general level of earnings in Ireland is determined. (See pp 2, 3). One point worthy of note in the above equation is that the coefficient of ΔW is significantly different from unity at the 5 per cent level. This could be construed as support for a milder form of the consolidation hypothesis, although Dicks-Mireaux and Shepherd, considering British experience in much greater detail than was done here, conclude that this is probably due to the fact that no proportional increase in 'supplementary payments' accompanied the increase in wage rates. On the whole these authors come down solidly against the consolidation hypothesis.

[11] L. A. Dicks-Mireaux and J. R. Shepherd, Unpublished paper quoted in J. C. R. Dow, *The Management of The British Economy 1945–60*, Columbia University Press 1965.

[12] A. G. Hines, 'Trade Unions and Wage Inflation in the United Kingdom 1893–1961', *Review of Economic Studies*, 31, (1964), 221–52.

[13] J. C. R. Dow, *op. cit.*, 354.

of aggregate hourly earnings than any other indicator of excess demand, aggregate or occupational. It was shown how entrepreneurial expectations concerning the future level of demand will lead to anticipatory hiring/dismissal, and that the magnitude of such anticipatory hiring will depend upon the rate of change of demand (approximated in this model by the rate of change of unemployment).

THE EMPIRICAL RESULTS

In the previous sections of this chapter certain relationships were postulated between the rate of change in average hourly earnings in transportable goods industries and excess demand in the key occupational labour market for electricians. In this section these relationships are tested.[14]

1954–1969

Taking the whole period under consideration the first test was on the basic relationship postulated in the previous section, i.e., $\Delta E_t = h(D_t^e)$ by regressing ΔE_t against the two components of D_t^e, the level of unemployment amongst electricians and the rate of change of that level.[15]

$$\Delta E_t = 13.31437 - 4.07149 U_t^e - 1.24002 \Delta U_{t-\frac{1}{4}}^e \quad R^2 = 0.5138;$$

$$D.W. = 2.58834 \quad (-3.48112) \quad (0.95404)$$

$$(t\text{–values in parentheses})$$

This equation clearly has reasonable predictive power and is significant at the 5 per cent level. The Durbin-Watson statistic indicates the absence of autocorrelation at the 5 per cent level. Regressing U_t^e on $\Delta U_{t-\frac{1}{4}}^e$ indicated the absence of multicollinearity at all levels of significance.

However, the hypothesis outlined earlier in this chapter makes a clear distinction between the relevance of the two explanatory variables, ΔU^e and U^e, at various levels of U^e. The hypothesis indicates that where U^e is relatively high we would expect

[14] All the data series used in this section are included in Appendix 3.
[15] The ΔU^e variable is lagged one quarter to allow for an 'initial delay period'—see L. A. Dicks-Mireaux, 'The Interrelationships between cost and price changes, 1946–59: a study of Inflation in Post-War Britain', *Oxford Economic Papers*, Vol. 13, 1961.

the more powerful explanatory variable to be U^e itself since little or no anticipatory hiring would be taking place and ΔU^e would therefore be an insignificant indicator of excess demand. Conversely where U^e is at or near its floor level anticipatory hiring would be expected to become the primary indicator of excess demand and ΔU^e to assume the dominant role in explaining variations in ΔE. Hence, in order to test the full hypothesis adequately it is necessary to examine the performance of each variable separately over different ranges of U^e. An examination of the statistics of unemployment amongst electricians (Appendix 2) reveals that the period 1954–1969 may be conveniently divided into two sub-periods according to the level of U^e which obtained. The first sub-period is the period 1954–1960 when the level of U^e ranged between an annual average maximum of 2.68 per cent and an annual average minimum of 1.57 per cent. During this sub-period then the level of excess demand for electricians was relatively low. The second sub-period is the period 1961–69 when the level of U^e ranged between an annual average maximum of 1.28 per cent and an annual average minimum of 0.81 per cent. During this sub-period the level of excess demand for electricians appears to have been relatively high. There is a clear distinction to be drawn between these two sub-periods in that the level of unemployment amongst electricians, and therefore the level of excess demand, appears to have shifted from one general level to another. According to the hypothesis U^e would be expected to be the more important explanatory variable in the first period and ΔU^e to be more important in the second.

1954–1960

For this sub-period ΔE was regressed against the single independent variable U^e; no time lags were assumed and the following result was obtained:

$$\Delta E_t = 15.25981 - 4.80474\ U_t^e;\ r^2 = 0.8725\ (\text{t-value in parentheses})$$
$$(-5.8310)$$

This result is striking in that a single independent variable, U^e, explains 87 per cent of the variation in ΔE_t. The inclusion of ΔU_t^e (or any lagged version of it) does not appreciably improve the fit and its coefficient is insignificant at both the 1 per cent and

5 per cent levels, indicating that anticipatory hiring did not occur to any extent during this sub-period. This result is, of course, consistent with the hypothesis.

1961–1969

Testing the hypothesis for the second sub-period required that ΔE_t should be regressed on $\Delta U^e_{t-\frac{1}{4}}$. A complication arose however: in 1965 wage rates were effectively frozen under the terms of a national wage agreement concluded in November 1964. Despite the fact that demand pressures on the wage level appear to have been extremely strong during the year, average earnings rose by only 3·1 per cent. Three courses were open to us here, (a) regress the equation using all the available data including that for 1965; (b) exclude 1965 data; and (c) include the 1965 data but include a dummy variable which would assume a value of unity for 1965 alone and a value of zero for all other years. We set out below the resultant equations for all three approaches.

(a) $\Delta E_t = 8.783 - 17.321\,\Delta U^e_{t-\frac{1}{4}};\quad r^2 = 0.646$
$\qquad\qquad (-3.575) \qquad\qquad\qquad$ (t–value in parentheses)

(b) $\Delta E_t = 9.4934 - 17.3619\,\Delta U^e_{t-\frac{1}{4}}\quad r^2 = 0.8364$
$\qquad\qquad (-5.53837) \qquad\qquad\quad$ (t–value in parentheses)

(c) $\Delta E_t = 9.49 - 17.36\,\Delta U^e_{t-\frac{1}{4}} - 6.39 D_t;\quad R^2 = 0.87$
$\qquad\qquad (-5.54) \qquad\quad (-3.27) \qquad$ (t–value in parentheses)

$\qquad$ (D_t = dummy variable to account for wage freeze year)

In all three equations $\Delta U^e_{t-\frac{1}{4}}$ is highly significant. The exclusion of 1965 data improves the fit considerably and the clear significance of the dummy variable reinforces our view that the rate of change of earnings in 1965 was severely reduced by the terms of the 1964 national wage agreement. This being the case this result is again striking in that a single independent variable, $\Delta U^e_{t-\frac{1}{4}}$ explains a high proportion of the variation in ΔE_t. The inclusion of U^e (variously lagged) into the equation did not improve the fit at all, indicating that during this sub-period U^e hovered around its floor level and the level of excess demand was reflected in changes in ΔU^e.

Testing for autocorrelation in the two sub-periods proved virtually meaningless. With only seven or eight observations one cannot justifiably accept or reject a null hypothesis of autocorrelation. However, an attempt to do so was made subject to this reservation.

There were two possible approaches to testing for autocorrelation:

(a) a d-value could have been generated and inferences made by extrapolation from the Durbin-Watson tables; or

(b) simply count the sign changes of the residuals and, assuming a binomial distribution, probability statements can then be made. Geary has shown by Monte Carlo methods that this means of testing for serial autocorrelation is (surprisingly) quite efficient relative to the standard Durbin-Watson test.[16]

The latter method was decided upon because theoretically it is more applicable on the case of a small sample. In both sub-periods the hypothesis of serial autocorrelation could not be accepted.

It has been observed how the two components of D^e perform almost completely independently of each other under different demand circumstances. The predictive power of the separate variables is therefore considerably greater within their respective ranges of dominance than that of the two variable model over the whole period. This raises the issue of how to use a model whose predictive capacity varies according to the magnitude of one of its independent variables. The answer to this question would be relatively simple of one could distinguish precisely between the ranges of U^e for which U^e dominates. This distinction was made theoretically earlier but only a guess is possible in numerical terms. It seems clear that no precise distinction between the two ranges of U^e can be made in practice since some area of overlap almost certainly exists. However, the implication of the choice of sub-periods is that the dividing line must lie somewhere in the range 1.2 per cent – 1.6 per cent U^e. Hence where $U^e > 1.6$ per cent it would be expected that U^e would be the primary explanatory variable and where $U^e < 1.2$ per cent ΔU^e would be expected to be the primary explanatory variable. This is admittedly imprecise but gives some impression of the boundaries of

[16] R. C. Geary, 'The Relative Efficiency of Count of Sign Changes for Assessing Residual Autoregression in Least Squares Regression', *Biometrika*, Volume 57, No. 1, April 1970.

the two ranges of U^e for which the separate influence of each variable dominates.

The predictive capacity of the two sub-models is dependent on the level of U^e at any given point of time. Clearly neither sub-model would be of much use if the level of U^e was fluctuating around the critical level which divides the two ranges from year to year. In such an event the whole period model would be required to predict changes in ΔE but would of course have a less powerful predictive capacity than either of the sub-models operating within their own ranges of U^e. In practice it appears that the level of U^e is likely to remain in the low range for the foreseeable future, unless policy measures or some other factors intervene. Hence the second sub-model with ΔU^e as the only independent variable will probably be the appropriate predictive model in the short term.

The main conclusion emerging from these findings is simply that they are consistent with the basic hypothesis outlined in earlier sections of this chapter. The rate of change of average hourly earnings in transportable goods industries in Ireland appears to be closely associated with the level of excess demand in the labour market for electricians. This is consistent with the view that electricians have, directly or indirectly, tended to determine the timing and magnitude of the key bargain which has in turn determined the rate at which the general wage level inflates.

POLICY IMPLICATIONS

The findings presented in the previous section lend support to the hypothesis that it is the level of excess demand in the labour market for electricians which determines the rate of change of the general wage level. These findings should not be interpreted in too narrow a manner because demand conditions for electricians may be a proxy for demand conditions for a group of skilled occupations. There is no way of discovering whether or not this is the case since data on occupational employment levels is not available in sufficient detail to conduct the relevant tests. Commonsense suggests however that, where electricians bargain within a larger group, e.g. in building and construction, and within the maintenance craft group, they will be in company with other groups of skilled labour for whose services there is an excess demand. In such cases the whole group is in a favourable position to lead movements in the general level of wages.

The important point here is that, although these results on data relate to electricians alone, it would be a mistake to believe that the rate of wage inflation experienced in Ireland is uniquely related to the demand situation obtaining for electricians. Excess demand undoubtedly exists, and has persisted for a number of years, for the services of a number of other skilled occupational groups in Ireland, notably mechanical fitters, welders, turners, toolmakers, motor mechanics etc. Even if the excess demand for electricians were to disappear over-night, it is unlikely that the course of wage inflation in Ireland would be in the least altered since some other group would spearhead the system with much the same force as electricians appear to do at present. The real problem therefore is that excess demand exists for the services of a whole range of skilled occupational groups within the Irish labour force and, so long as this situation persists, the rate of wage inflation is likely to be considerably more rapid than would be warranted by the conditions of aggregate demand for labour. The situation then is one on which an important, but numerically small, section of the Irish labour force experiences persistent excess demand while there is an overall excess supply of labour in the economy. The problem is obviously partly one of structural unemployment but it is also due to a fundamental factor disequilibrium characteristic of underdeveloped economies.[17] From the policy point of view these findings have relevance only to the problem of structural unemployment.

From the policy point of view the elimination of structural unemployment is clearly relevant to the problem of damping the rate of wage inflation. The wage mechanism is prevented from performing this function by the institutional characteristics of the system of wage determination in Ireland. Wages rise first in those labour markets in which there is an excess demand for labour and the size of the wage increase reflects the degree of that excess demand. If the wage structure responded everywhere to demand conditions specific to each occupation then a differential between the wages of those occupations for which there is an excess demand and those for which there is no excess demand would

[17] C. Mulvey, 'Employment Policy' in *Economic Policy in Ireland,* Dublin 1968.

emerge and a reallocation of labour from the latter to the former occupations could be expected. This reallocation would increase the supply of labour in those occupations where there is an excess demand and reduce the supply of labour in those occupations where there is an excess supply. In the long term the effect of this process would be to eliminate excess demand for labour in all areas of the labour market (although it would not proceed to the point of general full employment of labour) and the rate of change of wages would then reflect the general circumstances of demand for labour in the economy. However, because of the widespread application of comparability criteria in the Irish wage determination system the wage is prevented from playing its equilibrating role. The general level of wages rises at a rate which reflects the demand conditions of the tightest individual markets and it rises in a uniform manner. Hence, relative inter-occupational (and inter-industry) wage differentials remain fairly static over time and labour shortages and surpluses in different occupations persist.

The problem of dealing with this form of wage inflation therefore becomes a problem of economic policy. Two policy approaches suggest themselves:

(a) act to reduce the significance of comparability as a criterion of wage determination and allow the wage structure to allocate labour in an efficient manner, or,

(b) employ an active manpower policy to allocate labour in an efficient manner within the context of a rigid wage structure.

In both cases what is desired is a leftward shift of the Philip's curve. The first option attempts this by freeing the market of the institutional factors which stifle the allocative function of the wage mechanism; the second attempts this by acting directly on the structure of the labour supply. The first option is open to two important objections. The first is that it is impractical to attempt to persuade the trade unions to abandon a criterion which they believe (with justification in many cases) has served them admirably in the past and which, for many of them, is the only effective bargaining weapon open to them. The second objection is that even if the general abandonment of the comparability criterion could be secured, it is doubtful if the wage mechanism by itself would be capable of bringing about the necessary reallocation of labour.[18] The second option is considerably more attractive in that it can operate within the existing criteria of wage determination to which the trade unions are attached and it is also probably a more capable instrument of labour reallocation. Hence option (b) appears to be more realistic and also more likely to do an efficient job.

It would be tedious and unnecessary to discuss the general aspects of manpower policy directed towards eliminating shortages of skilled labour here. These have been widely discussed elsewhere (e.g. Davies[19]). The main priorities in the Irish case appear to be:

(a) to identify all of those occupations where labour shortages presently exist or are likely to emerge in the future (a precondition of such an identification would be the provision of a detailed occupational breakdown of the insured population by the Central Statistics Office on a quarterly basis);

(b) the provision of training and re-training facilities for new entrants to the labour force, suitable workers in low grade occupations and for suitable workers who are redundant for entry into these occupations;

(c) to emphasise training schemes rather than apprenticeship schemes as a means of acquiring a skill or range of skills;

(d) to obtain the co-operation of employers to encourage workers to acquire a skill by permitting day release and sandwich course release and to obtain the co-operation of trade unions to accept training as a means of acquiring a skill rather than apprenticeship;

(e) to provide suitable grants and incentives to attract sufficient numbers of workers to utilise the training and re-training schemes.

Even if these proposals were given immediate effect and were successful in attracting sufficient numbers of workers there would be a time lag varying from at least one to four years before their

---

[18] O.E.C.D., *Wages and Labour Mobility,* Paris 1965.
[19] G. Davies, 'Labour Supply Shortages: Causes, Consequences and Cures', *Journal of Economic Studies,* Volume 3, No. 1, (March 1968).

effects were felt in the tight markets. Hence, both from the point of view of economic efficiency and the particular problem of inflation these measures must be viewed as part of an essentially long term strategy.

There is, however, an interesting and rather unique dimension to the potential of manpower in Ireland. As a result of continuous heavy emigration of Irish labour to Britain there is a substantial pool of skilled Irish labour employed in Britain. It is likely that the majority of these emigrated to Britain as unskilled labour and acquired a skill in Britain. No doubt many of these skilled workers would wish to return to Ireland to work if they could be assured that they would suffer no loss of pay, inferior working conditions or inferior social conditions if they made the move. Already an Advisory Committee has been established to advise the Minister for Labour on the facilities which should be provided to emigrants wishing to return to Ireland to work. So far the only concrete result of this Committee's work is the publication and circulation amongst Irish workers in Britain of a regular bulletin listing job vacancies in Ireland.

It is clear that if this scheme developes it could yield a substantial inflow of skilled workers to the Irish labour market from Britain. If this were to be directed towards alleviating labour shortages in skilled occupations in the short run it would clearly be necessary for the Irish government to offer fairly substantial inducements to workers to return home. These might take the form of guaranteed housing, grants to cover removal costs, pay advances etc. However, it is clear that if a sufficient number of skilled workers could be induced to return to Ireland from Britain to appreciably reduce the rate of wage inflation in Ireland in the short term then the costs of the necessary inducements would probably be quite trivial compared to the savings. Hence there could be considerable merit in undertaking a crash programme to induce skilled Irish workers employed in Britain to return to Ireland even though such a programme might appear very costly in the short run. There are of course many imponderables in such a policy. It is likely that many Irish emigrants have become attached to the British way of life, and in particular, to the superior social welfare system, or have entered commitments which tie them to Britain in the short run and it is therefore by no means certain that even a fairly lavish system of inducements would be sufficient to attract the required numbers back to Ireland.

It is instructive to consider what are the implications of this analysis for prices and incomes policy, which is much discussed in relation to the problem of wage inflation in Ireland. The implications are in fact somewhat ambiguous. On the one hand the area of the labour market in which effective policy intervention would be necessary is clearly considerably smaller than is generally supposed. If prices and incomes policy could dampen the rate of wage inflation in the key occupations, then it could probably afford to neglect the remainder of the labour force and allow the criteria of comparability and relativity to extend the dampened rate of wage inflation into all other employments. However, this analysis implies that the source of inflationary pressure stems from excess demand in certain occupational labour markets and only takes on a cost push character after a key settlement or claim has occured. This process might merit the label 'demand induced cost inflation'. Incomes policy is generally inappropriate for dealing with demand inflation and would have great difficulty in attempting to dampen the rate of inflation in the key occupations. Prices and incomes policy might attempt to detach the demand induced inflationary markets from their cost push appendages in the rest of the labour force. However, this is a difficult task, as the N.B.P.I. discovered, and runs into the difficulties mentioned in option (a) above.

With or without incomes policy it is clear that a satisfactory long term solution to the problem of rapid wage inflation in Ireland depends primarily on an active manpower policy. The conventional longer term aspects of such a policy have been mentioned but special attention has been paid to the potential short term effects of a scheme designed to induce skilled Irish workers employed in Britain back to Ireland. Both of these policies, operating with short and long term perspectives, could contribute substantially to slackening the tight end of the labour market and reduce the rate of wage inflation as a result.

APPENDIX I

DEFINITION OF VARIABLES

ΔE_t = the annual percentage rate of change of average hourly earnings in transportable goods industries calculated from the third quarter of one year to the third quarter of the next, i.e.,

$$\Delta E_t = \frac{E_t - E_{t-1}}{E_{t-1}} \cdot \frac{100}{1},$$ where E_t is the third quarter index number of average hourly earnings in transportable goods industries; base 1953=100. Data from *Irish Statistical Bulletin*, various.[1,2]

ΔW_t = the annual percentage rate of change of average hourly wage rates for All Groups Dublin District, i.e.,

$$W_t = \frac{W_t - W_{t-1}}{W_{t-1}} \cdot \frac{100}{1},$$ where W_t is the annual index number of average hourly wage rates for All Groups Dublin District; base 1953=100. Data from *Statistical Abstract of Ireland*, various, Table 350.[2]

U_t^e = the annual average rate of unemployment amongst electricians and wiremen, including apprentices, calculated as the average percentage unemployment rate of the four quarters, 3rd Quarter $t-1$, 4th Quarter $t-1$, 1st Quarter t and 2nd Quarter t. Unemployment data is taken from the category Wiremen, Electricians in the Occupations of Persons on the Live Register, *Irish Statistical Bulletin*, various. Employment data is estimated as explained in Appendix 2.

$\Delta U_{t-\frac{1}{4}}^e$ = the annual change in the percentage of electricians unemployed calculated from the second quarter of $t-1$ to the second quatrer of t, i.e.

$\Delta U_{t-\frac{1}{4}}^e = (U^e\% \ 2nd \ Quarter \ t) - (U^e\% \ 2nd \ Quarter \ t-1)$ Data as for U^e above.[2]

U_t^a = the annual average level of unemployment measured from the average of the four quarters U_{t-1}^a (3), U_{t-1}^a (4), $U_t^a(1)$, $U_t^a(2)$. Numbers on Live Register (Non-

---

[1] This particular series was employed since it is the only series which appears to capture the full distributed effect of each wage round since 1954. Whether this is the result of chance or some seasonal pattern in the wage round system is not evident. Other ΔE series distribute the effects of each wage round over two year periods. Since the hypothesis concerns the initial conditions which detonate the round it was necessary to discover a ΔE series which reflected the cumulative effects of the initial detonation.

[2] Rate of change variables have not been centred since the hypothesis demands that distributed quarterly changes in these variables be captured in an annual series relating to the initial round (or non-round) conditions.

Agricultural) as a percentage of insured labour force.
Data from *Irish Statistical Bulletin,* various.
$\Delta U^a_{t-\frac{1}{4}} =$ Calculated as for $\Delta U^e_{t-\frac{1}{4}}$. Data as for U^a_t.[2]

APPENDIX II

ESTIMATES OF THE RATE OF UNEMPLOYMENT AMONGST ELECTRICIANS AND WIREMEN IN IRELAND 1951 TO 1969

There is no continuous statistical breakdown of the Irish labour force by occupation in any published source. The only such data available is supplied in the five yearly Census of Population published by the C.S.O. For intercensal years no estimates of the occupational composition of the labour force are made. After making strenuous efforts to obtain such estimates, from trade union membership figures, Department of Social Welfare, Department of Labour, etc. it was decided to use the following crude method of obtaining employment estimates for electricians and wiremen. Three census observations were available, for 1951, 1961 and 1966 (the occupational breakdown was not made in the 1956 census) and, after adjusting the 1951 data to the definitions employed in more recent census data, three figures for employed electricians and wiremen inclusive of apprentices were obtained. Estimates of the number of electricians employed in the building industry, obtained from the count of wet-time books exchanged in each year, for 1951, 1961 and 1966 were then calculated as a proportion of the totals obtained from the census for each year. The results were:

TABLE 1

	No. of Electricians employed in Building and Construction[3] (1)	No. of Electricians and Wiremen. Census data (2)	(1) as a % of (2) (3)
1951	1,319[1]	4,323[2]	30.5
1961	1,388	4,779	29.1
1966	2,151	7,084	30.0

[1] Figure for October 1950/October 1951. The accounting period for this data was altered to a January/December basis from 1953 onwards.

[2] This figure is adjusted to exclude 'other electrical fitters' since most of those so listed are categorised in later census breakdowns as telephone installers and are excluded from our definition.

[3] Source: *The Trend of Employment and Unemployment,* C.S.O. Table 6, various.

It will be noted that there is a remarkable stability in the proportion of electricians employed in building and construction to total electricians and wiremen in the three census years. Although it is recognised that this is a crude method of constructing employment estimates it is assumed that this stability obtained throughout the period 1951 to 1969. Hence the total number of electricians and wiremen in the labour force is estimated as 100/30 × the number of electricians employed annually in the building and construction industry. In defence of these estimates it may be said that unless this method is used no estimates can be made and also that for the purposes of this chapter, where the objective is to estimate rates of unemployment, there is room for some margin of error in the denominator without distorting the results. Hence the following employment estimates:

TABLE 2

	No. of Electricians Employed in Building and Construction[1]	Estimated No. of Electricians and Wiremen in Labour force	Census
	(I)	(I) × 100/30	
1951	1,319	4,397	4,323
52	1,397	4,656	
53	1,423	4,743	
54	1,482	4,940	
55	1,512	5,040	
56	1,403	4,676	
57	1,512	5,040	
58	1,507	5,023	
59	1,448	4,826	
60	1,399	4,663	
61	1,388	4,626	4,779
62	1,505	5,016	
63	1,622	5,406	
64	1,785	5,950	
65	1,855	6,183	
66	2,151	7,170	7,084
67	2,196	7,320	
68	2,354	7,846	
69	2,521	8,403	

[1] Source: *The Trend of Employment and Unemployment,* C.S.O. Table 6, various issues.

The next step in calculating the rate of unemployment amongst electricians and wiremen was to collect the data published quarterly in the occupational breakdown of the live register published by the C.S.O. in the *Irish Statistical Bulletin.* The data was collected quarterly on the following basis:

1st Quarter: April; 2nd Quarter: July; 3rd Quarter: October; 4th Quarter: January of following year. The results are set out over:

TABLE 3

Estimated Unemployment rate amongst Electricians and Wiremen. Quarterly Averages 1953 to 1969

Year & Q.	No. Unemployed[1]	Unemployment Rate[2] %	Year & Q.	No. Unemployed	Unemployment Rate %
1953 (1)	155	3.27	1963 (1)	66	1.22
(2)	156	3.27	(2)	66	1.22
(3)	110	2.32	(3)	49	0.90
(4)	88	1.85	(4)	38	0.70
1954 (1)	124	2.51	1964 (1)	63	1.05
(2)	128	2.60	(2)	52	0.87
(3)	77	1.56	(3)	76	1.27
(4)	87	1.76	(4)	61	1.02
1955 (1)	98	1.94	1965 (1)	50	0.80
(2)	98	1.94	(2)	54	0.87
(3)	60	1.19	(3)	45	0.72
(4)	61	1.21	(4)	54	0.87
1956 (1)	81	1.73	1966 (1)	63	0.87
(2)	103	2.20	(2)	63	0.87
(3)	103	2.20	(3)	51	0.71
(4)	152	3.25	(4)	58	0.80
1957 (1)	173	3.43	1967 (1)	70	0.95
(2)	148	2.93	(2)	81	1.10
(3)	114	2.26	(3)	71	0.96
(4)	105	2.08	(4)	84	1.14
1958 (1)	125	2.48	1968 (1)	121	1.42
(2)	125	2.48	(2)	92	1.17
(3)	107	2.13	(3)	103	1.31
(4)	101	2.01	(4)	95	1.21
1959 (1)	97	2.01	1969 (1)	95	1.13
(2)	142	2.94	(2)	79	0.95
(3)	74	1.53	(3)	79	0.94
(4)	71	1.47	(4)	87	1.03
1960 (1)	67	1.43			
(2)	43	0.92			
(3)	56	1.20			
(4)	43	0.92			
1961 (1)	49	1.05			
(2)	51	1.10			
(3)	43	0.92			
(4)	45	0.97			
1962 (1)	50	0.99			
(2)	49	0.97			
(3)	47	0.94			
(4)	43	0.86			

[1] Source: *Irish Statistical Bulletin,* Occupational Breakdown of Unemployment, Electricians and Wiremen, various issues.

[2] Number Unemployed as a percentage of estimated number of Electricians and Wiremen in Labour Force (See Table 2) in each year.

16

APPENDIX III

DATA

	ΔE_t	U_t^e	$\Delta U_{t-\frac{1}{2}}^e$	U_t^a	ΔU_t^a	$U_t^{an\ av}$
1954	1.9	2.5	−0.67	7.525	−2.1	7.8
1955	5.6	1.96	−0.66	6.400	−1.2	6.4
1956	8.0	1.51	+0.26	8.450	+0.6	7.8
1957	1.6	2.77	+0.73	8.50	+1.6	8.7
1958	4.5	2.44	−0.55	8.25	−0.4	8.6
1959	2.8	2.76	+0.56	7.325	−0.9	8.0
1960	7.2	1.84	−2.02	5.825	−1.5	6.7
1961	5.8	1.02	+0.18	5.5	−0.6	5.7
1962	13.1	0.99	−0.13	5.7	−0.1	5.7
1963	3.3	1.00	+0.25	5.625	+0.1	6.1
1964	13.0	0.97	−0.35	5.45	−0.3	5.7
1965	3.1	0.99	0.00	5.4	−0.2	5.6
1966	11.6	0.83	0.00	6.175	−0.9	6.1
1967	4.8	0.83	+0.23	6.525	+0.2	6.5
1968	10.0	1.15	+0.07	6.13	+0.1	6.5
1969	14.0	1.01	−0.23	5.8	−0.4	5.8†

† = estimate.
See Appendix 1 for definition of variables and sources of data.

APPENDIX IV

In order to compare the explanatory power of this model to that of other possible models, the relation between ΔE_t and the more conventional indicators of aggregate excess demand was examined. The results of these regressions are given below; also included below are those regressions which contained a variable from the labour market for electricians as an independent variable, but whose performance was less satisfactory than those equations included in the findings section. Student-t-values are included in parentheses below the coefficient.

1954–1969

(1) $\Delta E_t = 13.40466 - 4.00637 U_t^e$ $r^2 = 0.4769$
                  (-3.44321)

(2) $\Delta E_t = 6.95503 - 0.97614 \Delta U_{t-\frac{1}{2}}^e$ $r^2 = 0.0229$
                  (-0.55223)

(3) $\Delta E_t = 23.18327 - 2.37696 U_t^a + 1.35072 \Delta U_t^a$ $R^2 = 0.3851$
              (-2.64656) (1.28080)

(4) $\Delta E_t = 21.14362 - 2.12295 U_t^a$ $r^2 = 0.3010$
      (-2.36598)

(5) $\Delta E_t = 7.30251 + 0.73443\Delta U_t^a$ $r^2 = 0.026$
      (0.59056)

1954–1960

(1) $\Delta E_t = 15.01974 - 4.71394 U_t^e - 0.10542\Delta U_{t-\frac{1}{2}}^e$ $R^2 = 0.8735$
      (-4.61920) (-0.20069)

(2) $\Delta E_t = 4.03207 - 1.18104\Delta U_{t-\frac{1}{2}}^e$ $r^2 = 0.1989$
      (-1.11425)

(3) $\Delta E_t = 14.46463 - 1.29176 U_t^a + 0.69892\Delta U_t^a$ $R^2 = 0.1425$
      (-0.81361) (0.54081)

(4) $\Delta E_t = 9.51580 - 0.68122 U_t^a$ $r^2 = 0.0798$
      (-0.65860)

(5) $\Delta E_t = 4.40141 - 0.04875\Delta U_t^a$ $r^2 = 0$
      (-0.05556)

1961–1969

(1) $\Delta E_t = 8.51288 + 1.00555 U_t^e - 17.31819\Delta U_t^e$ $R^2 = 0.837$
      (0.13784) (-5.03116)

(2) $\Delta E_t = 5.17168 + 4.38801 U_t^e$ $r^2 = 0.012$
      (0.26875)

(3) $\Delta E_t = 24.8588 - 2.62596 U_t^a + 0.99679\Delta U_t^a$ $R^2 = 0.034$
      (-0.40345) (0.18720)

(4) $\Delta E_t = 20.15804 - 1.82684 U_t^a$ $r^2 = 0.027$
      (-0.40596)

(5) $\Delta E_t = 9.44483 - 0.41310\Delta U_t^a$ $r^2 = 0.002$
      (-0.11084)

O'Herlihy's model of earnings inflation was tested after additional observations had been obtained and the following was the resultant relation.

$$\Delta E_t = 12.11403 - 0.31582\overline{U}_t + 1.01471\Delta P_t \qquad R^2 = 0.4866$$
$$(-2.0250) \qquad (2.62619)$$

t-values in parentheses.
period : 1954–1968

This equation did not perform very well when subjected to the standard statistical tests of significance.

The inclusion of the profits change variable (Cowling) improved the fit significantly, the resultant equation being:[20]

$$\Delta E_t = 5.86439 - 0.18093\overline{U}_t + 1.23503\Delta P'_t + 0.25956\Delta \overline{D}_t;$$
$$(-1.43188) \quad (4.08779) \quad (3.01767)$$
$$R^2 = 0.7313$$
$$\text{period:}$$
$$1954-1968.$$

For the definition of the variables, see Cowling.[21]

Cowling's model was tested for the two sub-periods 1954–1960 and 1961–1968, but the coefficients and the R^2s turned out (as would be expected) to be insignificant due to the small number of degrees of freedom. Excluded from this test of the Cowling model are his unionisation series, for two reasons:

(a) no reliance was placed on the accuracy of the unionisation series;

(b) the coefficient of the unionisation variable was insignificant in Cowling's original model.

---

[20] See the reservations in C. Mulvey and J. Trevithick, 'Wage Inflation: Causes and Cures', *Quarterly Bulletin of the Central Bank of Ireland*, Winter, 1970, 112.

[21] K. Cowling, *op. cit.*

INDEX

Aer Lingus, 51–2
Agricultural property, 162, 173; *see also* Undervaluation
Agricultural sector of the economy, 85
Agricultural subsidies, 51
Allowances against income tax, 82; *see also* Income taxation
Angels: economists' estimates on, 63
Assets portfolios, 3
Assets at death *see* Undervaluation
Australia's national debt, 45
Austria's national debt, 45
Automatic stabilisers and the budget, 13; *see also* Built-in flexibility
Average cost pricing, 103

Balanced current budget *see* Current budget
Balance of payments, 6, 17, 31
Balopoulos, E.T., 71, 74, 85
Banks and purchase of debt, 60
Bargaining and wage leadership, 207, 208
Beef: principal export, 125–6; demand curve, 143–4; Argentinian, 148
Beer as principal export, 126
Belgium's national debt, 45
Birth rates, 187
Bond policy and an orderly market, 59–60
Bord na Mona: and loans, 40; and the E.S.B., 102; and employment, 118
Borrowing: by the government, 3–5; residual, 31; from abroad in relation to full employment, 55; as a policy issue, 57; *see also* Debt *and* Government borrowing
Bottlenecks and supply of skilled personnel, 199, 211
Bovine tuberculosis, 48
Brain drain: cost of, 202
British import levy and deposits, 19
Budget:
    1953–54, 26
    1960–61, 16, 25
    1961–62, 24, 25
    1963–64, 16, 17, 24
    1966–67, 17
    1967–68, 16, 18, 25
    1968–69, 12, 17
    1969–70, 17
    1970–71, 16
Budgets: planned, 12; deficits unplanned, 17; *see also* Capital budget *and* Current budget
Building and wage leadership, 208
Built-in flexibility: and the current budget, 13; and corporate income taxation, 68–9, 88–95; and surtax, 81; and agricultural sector, 85
Butter: principal export, 125–6; supply function of, 144

Canada's national debt, 45
Capital: and current government expenditures—irrelevant distinction, 6; from new issues of securities, 53; raised externally and multiplier effect, 65
Capital budget: 19–32, 35; and growth, 24
Capital gains, 60
Capital inflows, 31
Capital market underdeveloped, 54
Cash proportion in net wealth, 173
Cash ratio, 3, 60, 61
Cattle as principal export, 125–6
Central Bank: 22, 29, 30, 124; and exceptional finance, 30
Ceylon's national debt, 46
C.I.E. flotations, 51
Commissioners of Inland Revenue and annual estimate of personal wealth, 160
Companies: income tax on, 90
Comparability criteria, 206, 224
Construction and wage leadership, 208
Consumers' attitudes and tastes, 135
Consumption: lifetime, and brain drain, 189
Corporate income taxation *see* Built-in flexibility
Corporate sector, 68–9; *see also* Built-in flexibility
Cost inflation: demand induced, 227
Cost-minimising, 106
Cowling, K., 204
Cross-elasticities, 128, 129, 152
Cross-subsidisation: rural and urban, 114, 115
Current account surpluses, 18

Current and capital expenditures: blurred distinction between, 18, 48

Current balance of payments: deficits, 31

Current budget: 9–12; balanced, 12, 16–17, 32–5; deficit, 34

Customs and excise taxes, 95

Day release, 225

Debt: redemption, 19, 47; short-term, 41; service cost of, 42, 64; purchase, 46; Irish banks' attitude to purchase, 49; new and old, 54; and money supply, 57; sale of, 57 see also Deflation; government official holdings of, 62; size constraints on, 63; viewed as asset creation, 63

Deflation: and debt sales, 57, 62; and falling bond prices, 59

Demand for electricians, see Electricians and Labour

Demand management: 26, 41; and fiscal policy, 32

Denmark's butter and beer, 148

Devaluation: of the Irish pound, 124, 125; unilateral Irish, 154

Differentials and inter-industry wage-structure, 207

Discounted cash flow and external capital, 65

Discount rate and external debt issuance, 65

Distributional effect of electricity charges, 120

Dividends, 68

Eckstein, O., 208

Economic forecast, 6, 37

Education: 15, 48; of emigrants, 190; higher, and externalities, 190–92; policies and brain drain, 198; data on qualifications, 201

Elasticity; of taxes, 95; of substitution, 140 see also Import demand elasticities

Elderly people, 176

Elections and role of debt, 46

Electrical contracting and wage leadership, 208

Electrical fitters see Electricians

Electricians: wage leadership and unemployment rate, 208; excess demand for, 209, 222; and general

demand for labour; 210; wage bargain criterion for economy, 215

Electricity: rural versus urban consumers, 102; turf-fired stations, 102, 103, 105–13; oil versus turf, 102, 103; coal versus oil, 106; fuel costs, 106; capital costs, 109; excess costs, 110; locational problems, 111; rural deficits, 114

Electricity Supply Board: flotations, 51; overheads and production efficiency, 107; see also Electricity

Emigrants: remittances, 189; state payment of fares, 195, 196; Irish to U.S., 200

Emigration and unemployment, 195

Employment see Full employment

Equities: ratio to bonds, 60

Estate duty and calculation of property ownership, 160; see also Evasion

Estates: large and small and their treatment, 161–6, 179–85; and presentation for probate, 162; age and sex of owners, 166

Evasion of estate duty, 161

Exchange rate: changes in, 124

Exchequer notes, 39

Export: elasticity of foreign demand, 124; changes in supply and demand, 127; exchange elasticity of receipts, 129; demand and its price elasticity, 130, 154; supply and its price elasticity, 136–8

External assets, 61

External liabilities and national debt policy, 52

Family unit and migration, 193

Food and Agricultural Organisation, 146

Farm incomes, 16

Federated Union of Employers, 205

Fiscal policy: and state of economy, 6; philosophy of, 8; risk for developing country, 26; not consistent with monetary policy, 36

Floating debt, 41

France: its national debt, 45; and beef consumption, 143

Fuel costs see Electricity

Full employment: 8, 32; and new debt, 55

Funding and the national debt, 41–2

Germany: its national debt, 45; and beer, 148

Gifts *inter vivos,* 161, 178

Government: decrees and subsidies, 102; savings on services of, 202

Government borrowing: 35; external, 22–3, 27, 33, 40, 41, 42; from banks, 22–3, 27, 33; from non-bank public, 22–3; guaranteed loans, 40; improved marketability of bonds, 42; taken up by associated banks, 49, 50

Governments and elections, 46

Grants and incentives for training workers, 225

Grubel, H., 188, 192, 193, 201, 202

Haughey, C. E., 16, 18, 25, 26

Hart, J., 146

Health, 15

Homogeneous products and countries of origin, 134

Horner, F. B., 132, 149

Housing: 2; and subsidies, 51

Houthakker, H.S., 139

Hydro electricity, 112

Import demand elasticities, 138–46

Incentive to work, 50

Income: distribution of, 71, 99; redistribution and electricity, 122; from investments abroad undervalued, 174

Income-elasticity of demand for electricity, 121

Income taxation: personal, 68; allowances for children, 68, 82; marginal rate of personal, 71, 78; and electricity, 119

Indirect taxes *see* Taxes, indirect

Inflation: 34; contained, 17; government interest in continuing, 64

Insurance and prices, 131

Interest rates, 3

Internal assets, 61

International Development Association 19

International Monetary Fund, 19, 24, 30, 148

Investment: 35, 50; private fixed, 6

Kaldor, N., and welfare compensation, 116

Labour: demand for, 210; prediction of demand, 212; shortages, 225; pool of skilled in Britain, 226

Land bonds, 40

Leser, C.E.V., 146

Life insurance and estates at death, 162

Liquidity: ratio, 3, 60, 61; leakage, 58

Loans: repayment, 22; guaranteed by the State, 40

Local authority finance, 51

Locked-in, 58

Lydall, H. F., 160, 175, 182

Lynch, J., 17, 26, 28, 29

Lynn, R., 186, 201–3

McAleese, D., 95, 96, 139

McDowell, M., 89, 95

MacEntee, S., 26

Magee. S. P., 139

Maintenance craftsmen and wage leadership, 208

Manpower, 227

Marginal cost pricing: subsidies and deficits, 103

Marketable securities: 39; and new capital, 53

Market share scale factors, 150

Marshall's 'law of demand', 125

Mechanical fitters: excess demand for, 223

Migration *see* Emigration

Money supply: 4; and debt, 57; *see also* Debt

Monopolist and subsidies, 99

Motor mechanics: excess demand for, 223

Murphy, D. C., 146

Musgrave, R.A., 58

National debt: maturity of, 39–41; cost of servicing, 42; increasing size of, 43; interest charge, 43; and G.N.P., 43–4; American and British influences, 44–5; and financing capital schemes, 47; and 'new-orthodoxy' controversy, 61; *see also* Debt.

National Development Fund, 22

Nevin, E. T., 160, 163

New Zealand's national debt, 45

Norway's national debt, 45

Objectives of economic policy, 1

O'Herlihy, C. St J., 204, 205

O'Mahony, D., 205
Open market operation, 62
Organisation for Economic Co-operation and Development, 36, 143, 150

Pay-as-you-Earn, 16
Post Office Savings Bank, 22, 28
Pratschke, J. L., 121, 123, 146
Prest, A. R., 74
Price elasticity: 143; of electricity demand, 120; of export supply and demand, 125; of supply and demand for U.K. imports of agricultural products, 145; of total domestic output, 146-7
Prices: stability, 6, 8, 32; and subsidies, 99; changes and electricity, 119; indices, 125; and incomes policy, 227
Private motorists, 101
Private sector and public sector competing for funds, 52
Productive capacity and debt financing 56
Profit maximisation, 99, 103-4
Profits, 8, 68; and tax yield, 89; tax rates, 94; normal and average cost pricing, 103-4
Progressive tax structure, 191
Property: ownership of, 159-78
Public capital expenditure: 31; White Paper on, 25
Public sector wages, 15

Quota, 124

Radcliffe Report, 59, 61
Rate of personal tax see Income taxation
Rates: 51; exemption of Electricity Supply Board, 103, 115-16
Reason, L., 84, 85
Recycling debt, 41
Redistribution and subsidies, 100
Regressiveness and electricity pricing, 121
Rents, 68
Revenue: automatic changes, 13-15; discretionary influences, 13-15
Revenue Commissioners and perfect foresight, 75
Revolutions: intellectual involvement in, 190

Risk taking, 50
Ryan, Dr J., 16, 17, 24

Sandwich courses, 225
Savings: 2, 29, 39, 49, 50; division between public and private sectors, 52; invested outside Ireland, 54; transfer between assets, 54
Scitovsky, T., and welfare compensation, 116
Scott, A., 188, 192, 193, 201, 202
Second Programme for Economic Expansion, 33, 37
Second-best theory, 117
Semi-state bodies: 51; and debt, 40
Short-term debt see Debt
Sinking funds: 35, 42; an anachronism, 43; provision by Electricity Supply Board, 105
Social Insurance Fund, 22
Social welfare payments, 15
State-guaranteed borrowing, 24; see also Debt
Sterling/Deutschemark Loan, 30
Stock exchange securities: probate evaluation of, 162
Stocks and shares: foreign, held by Irishmen, 173; Irish, and personal wealth, 173; undervalued for probate, 174
Stone, R., 141
Structural deficiencies and debt, 56
Subsidies: as negative taxes, 98; definition of, 98, 101; category of, 99; and income distribution, 99; cross, 102, 103
Surtax as contribution to built-in flexibility, 81
Sweden's national debt, 45

Tariffs: for electricity peak and base load, 104; new, 124; and prices, 131
Taxes: 2; indirect, 2, 15; direct, 15; and dis-utility, 47; and debt interest payments, 47; changes related to personal income and companies, 72-3; rates and changes in income, 75; earmarked, 101; losses from emigration, 202; see also Income taxation
Taylor, L. D., 139
Technical progress, 183
Third-party insurance for motorists as subsidies, 101

Third Programme for Economic and Social Development, 37

Thomas, D., 193

Time lag between death and probate presentation, 181

Tipping, D. G., 160, 175, 182

Toolmakers: excess demand for, 223

Trade-offs between government objectives, 7

Trade Union Congress, 205

Trade unions comparability criterion, 206

Training facilities, 225

Training schemes versus apprenticeship, 225

Transfer payments as subsidies, 98

Transport costs and balance of payments, 124; and prices, 131

Treasury Bill supply: 39, 'new-orthodoxy' controversy, 62

Turnover tax, 95

Turnovsky, S. J., 139

Undervaluation of estates for probate, 162, 173

Unemployment: short-term, 194; long term or structural, 195

Unionisation: rate of, 205

United Kingdom's national debt, 45

Unit fuel costs, 102

Unit of output and subsidies, 99

U.S.A.'s national debt, 45

Utility: of the individual, 46; consumers', and subsidies, 100

Velocity of circulation, 4

Wage: marginal rates of taxation of, 79; rounds, 205; settlements uniform, 206

Wage leadership: 207; and industry, 208; occupational, 208; and skill, 208

Walsh, B., 146, 148

War and debt, 47

Ways and means, 39, 42

Wealth: Ireland compared to Great Britain, 169–71; components of personal, 171–2

Wealth effect, 3, 54, 55, 58, 63

Weisbrod, B., 195

Welders: excess demand for, 223

Welfare loss and electricity consumers, 120

Wilson, T. A., 208

Workers' retraining, 48

World Bank, 19